The Mineral Position of the
United States, 1975–2000

Proceedings of a Symposium Sponsored by the

Society of Economic Geologists

At Minneapolis, Minnesota, November 1972

The Mineral Position of the United States, 1975-2000

Edited by
Eugene N. Cameron

Published for the
Society of Economic Geologists Foundation, Inc.
by
The University of Wisconsin Press

Published 1973
The University of Wisconsin Press
Box 1379, Madison, Wisconsin 53701

The University of Wisconsin Press, Ltd.
70 Great Russell Street, London

Printings 1973, 1974

Printed in the United States of America

ISBN 0-299-06300-3 cloth. 0-299-06304-6 paper
LC 72-7983

Contents

v

Figures

vii

Tables

Preface

The symposium at which the papers composing this book were presented was held at Minneapolis, Minnesota, on November 15, 1972, during the joint annual meeting of the Geological Society of America and Associated Societies, of which the Society of Economic Geologists is one. The symposium was sponsored by the Society of Economic Geologists and moderated by Elburt F. Osborn, President of the Society. Publication has been made possible by a grant from the Society of Economic Geologists Foundation, Inc.

A symposium is made possible by the efforts of many individuals. Members of various program committees of the Society of Economic Geologists, especially Stewart R. Wallace (chairman, 1971–72), Ulrich B. Peterson (chairman, 1972–73), and Robert B. Hoy, contributed greatly to development of the program and to the attendant arrangements, as did discussions held during 1971 and 1972 with the Program Policy Committee and Council of the Society. At Minneapolis, George R. Rapp, Jr., David Southwick, and Ernest K. Lehmann, all of the local committee of the Geological Society, were always most helpful, not to say indispensable. The thanks of the Society of Economic Geologists are due all these people.

Finally, I wish to salute my fellow authors for their cooperation in meeting deadlines set for preparation and presentation of these papers. Special appreciation is due to Herbert I. Fusfeld, Director of Research, Kennecott Copper Corporation, who, when David Swan was unavoidably detained in New York, ably presented the paper, "Science and Technology—Aid and Hindrance."

Eugene N. Cameron

Madison, Wisconsin
December 1972

Introduction

National problems associated with mineral supply and demand are moving to center stage, as foreseen in the conclusions of the twenty-year-old report of the President's Materials Policy Commission. In the past few years a serious energy problem has developed, a worsening balance of trade continues, added concern has been directed to occupational health and safety, and the new and unpredicted element, "environment," threatens to engulf the mineral industries. All this in the face of a declining domestic capability to supply needed primary minerals.

The federal government is reacting. The National Mining and Minerals Policy Act of 1970 states that

The Congress declares that it is the continuing policy of the Federal government in the national interest to foster and encourage private enterprise in (1) the development of economically sound and stable domestic mining, minerals, metal and mineral reclamation industries, (2) the orderly and economic development of domestic mineral resources, reserves, and reclamation of metals and minerals to help assure satisfaction of industrial, security and environmental needs, (3) mining, mineral, and metallurgical research, including the use and recycling of scrap to promote the wise and efficient use of our natural and reclaimable mineral resources, and (4) the study and development of methods for the disposal, control, and reclamation of mineral waste products, and the reclamation of mined land, so as to lessen any adverse impact of mineral extraction and processing upon the physical environment that may result from mining or mineral activities. For the purpose of this Act "minerals" shall include all minerals and mineral fuels including oil, gas, coal, oil shale and uranium.

In an amendment to this act, the Congress in 1972 passed legislation providing annual funding on a matching basis to support a mineral science and engineering center at one state-tax-supported educational institution in each participating state—an action to provide support in the mineral resources field similar to the present support of agricultural resources education and research. Although the bill was not enacted into law, the increasing governmental concern about the present state of mineral technology is evident from this congressional action.

In another move by the federal government, the National Commission on Materials Policy was established through provisions in Title II of the Resource Recovery Act of 1970. The commission was directed to "make a full and complete investigation and study for the purpose of developing a national materials policy . . . ." The commission is now hard at work on the report, which must be submitted to the president and to the Congress no later than June 30, 1973.

In view of these developments the Council of the Society of Economic Geologists decided that a review of the national mineral situation for the geological profession and for the public would be timely and highly appropriate. Professor E. N. Cameron, President-elect of the society, was given the assignment of developing a program. This evolved into the symposium on "The Mineral Position of the United States, 1975–2000," held as part of the annual meeting in Minneapolis in November 1972. The views of eight distinguished persons representing various branches of the mineral field and from government, industry, and educational institutions were presented. These highly relevant and significant papers are contained in this volume.

Elburt F. Osborn
President
Society of Economic Geologists

December 1972

Contributors

JAMES BOYD
President, Materials Associates, Washington, D.C.

EUGENE N. CAMERON
Van Hise Distinguished Professor of Geology, Department of Geology and Geophysics, University of Wisconsin, Madison

STANLEY DEMPSEY
General Attorney, Law Department, Western Area, American Metal Climax, Inc., Denver, Colorado

PETER T. FLAWN
President, The University of Texas at San Antonio

VINCENT E. MCKELVEY
Director, U.S. Geological Survey, Washington, D.C.

JOHN D. MORGAN, JR.
Assistant Director, Mineral Position Analysis, Bureau of Mines, Department of the Interior, Washington, D.C.

JOHN DREW RIDGE
Professor of Economic Geology and Mineral Economics, Department of Mineral Economics, The Pennsylvania State University, State College, Pennsylvania

DAVID SWAN
Vice President, Technology, Kennecott Copper Corporation, New York, New York

1

James Boyd

Minerals and How We Use Them

Minerals make it possible for man to design products in materials having the properties required to perform certain specific functions. The products of the forests, the farms, the seas, the air, and the earth's crust provide man with those materials required for his multitude of needs. Minerals come primarily from the earth's crust, but the products of erosion of that crust are sometimes sufficiently concentrated in the seas and become an economic source.

When man desires to create a product for his use, he searches for that material which will best provide the properties required of that product. In doing so, he usually gives due consideration to availability and cost. In the future he must also consider the ease of returnability to the material cycle.

Realization of this concept of material use is vital to appraisal of the world's material posture. The use of the term *material* was made advisedly, as the interchangeability of materials to perform a function is becoming a major consideration in product design. There are many ways in which "minerals" can be used interchangeably with forest, farm, or synthetic products to produce comparable properties.

As an illustration, until World War I, nitrogen was a mineral product. The changing availability of nitrates forced industry to apply existing scientific knowledge to produce nitrogen from the air. Even when a mineral, natural gas, is used to produce ammonia, the source of nitrogen is still from the atmosphere.

As the specific assignment here is *minerals*, this chapter will confine itself to that aspect of material supply.

Energy is uppermost in the public's mind. It is well, then, to start with that subject. Most of the world's currently used energy materials obtain their energy from the sun. As that solar energy was captured by plants and stored as fossil fuels, it entered the crust of the earth and became a mineral (coal, liquid petroleum, gas, oil, shale, and tar sands). Only as hydroelectric power is energy derived directly from the sun in any degree of magnitude. And this will remain so until man learns to harness the sun's direct rays economically in substantial quantity.

As he substitutes atomic energy for solar or fossilized solar energy, man will stay with mineral materials in the form of uranium, thorium, or deuterium from the seas (the subtlety of the latter classification is unimportant to the discussion). It is important to remember that man's need for a property is *energy*, not coal, oil, gas, or electricity. It is important, however, to provide that energy in a form required by the function he wants energy to perform, for example, stationary or motive power. Looked at in this light, the availability of energy materials is so enormous that the possibility of running out of energy materials in any conceivable time-frame is preposterous. Man can, and has, already involved himself in a dilemma. By failing to use his ingenuity to develop the proper sources of energy materials to meet his needs, he has created the impression that energy availability is finite.

Others in this book will tackle this problem in more detail. It is mentioned here mainly to illustrate, with the vividness of current events, that the supply of properties demanded of minerals can come from many sources.

The world's demand for food stuffs grows with population gains and per capita consumption. As its remaining arable lands become poorer and more inaccessible, these demands can be met only by improved productivity. Greater productivity is achieved by biological means and the provision of nutrients deficient in the land. The latter depends on minerals, for plants have, since plant life began, depended on the release of chemicals from minerals into the soil. This is not to ignore the sources of carbon and nitrogen from the air. As land is more intensively cultivated, such nutritive elements as potassium, phosphorus, and some trace elements must be provided from other sources. These sources are usually not from the common rock similar to those which provided the original supply. Nature by geologic processes has concentrated these elements in deposits which can be exploited cheaply. As long as they last, and the resources are enormous, it will be unnecessary for man to go back to obtaining these supplies from the limitless primary rock.

Man has many more needs and wants than energy and food. From the stone axe to the atomic bomb, he has depended largely on minerals for his weapons. Even as his weapons have become more sophisticated, so has his need for mineral materials demanded more complicated properties and, therefore, a greater variety of minerals. Alley Oop's contemporaries needed only common rock of almost any kind. These rocks would still be in ample supply even if we still used them as weapons.

Modern man needs more than ninety mineral materials produced from the earth to provide him with the properties he needs. He can, however, obtain similar if not comparable properties from more than one element or mineral. There are many properties he would like to have in his materials which his ingenuity has thus far failed to produce, for example, materials of sufficient strength at high temperatures to make it possible to use his engines more efficiently. It is improbable that the geologist will be much help in reaching for these goals. It is more likely that the physical scientist will find the means to treat those minerals already made available in the way necessary to produce the desired result.

But man, in satisfying his everyday needs and wants, is entirely dependent on materials. Although he uses more materials on a tonnage basis from the forests and from the farms for nonfood purposes, he could not use even these materials without minerals. The iron to cut the trees into lumber and to fasten the boards together are examples.

Most people remain blissfully unaware of their complete dependence on almost the whole gamut of functions required of mineral properties in their every activity—from the metal that brings utilities in and distributes them by wire to the pipe that removes his wastes by conduits made of clay or iron or mineral-derived plastics. The concerned person would be amazed if he knew how many people were employed in providing him the innumerable mineral materials required to bring about the elaborate systems in his television, telephone, and refrigerator. He would be fascinated by the complexity of the effort required by the geologist just to find the deposits of minerals needed to provide the properties required to make his transportation system operate reliably at the speed and convenience he demands. Each year twenty-three million tons of ferrous and non-ferrous mineral materials are required to build automobiles.

Thus, man uses in one form or another a major portion of the 103 elements known to him, most of which come to him from the earth in the form of minerals. Without them and the complicated technology associated with them, civilized man would still be striving to rise above the level of mere subsistence, and those in lesser developed countries would have no hope of ever doing so.

The public is constantly being told that they are being faced with the exhaustion of the sources of their materials supply. How true is this?

In talking about the interchangeability of the materials available to impart the required properties, we are in fact assuring supplies of properties in such magnitude that they cannot run out in any imaginable period. It is important that the geological disciplines are prepared to articulate this principle without underestimating the magnitude of the problems of supply.

The National Materials Policy Commission has asked the Mining and Metallurgical Society of America to define *resources* in the terms I have used above. We have asked the Society of Economic Geologists, the American Institute of Professional Geologists, and the American Institute of Mining, Metallurgical, and Petroleum Engineers to define what we mean by *reserves*, and those definitions recommended by a joint committee are now before the Council of the Society of Economic Geologists.

It is essential to make it clear to the public that total resources are large enough to stagger the imagination; that reserves are those resources which have been delineated as viable in the context of the current economic and technological conditions; and that because of this definition and the cost of finding and delineation, reserves will always appear to be limited. The public thinks of published reserves as supply. But supply is that volume of materials that can be made available through the industrial complex built on known reserves. Shortages, if they occur, result from the failure to find and develop reserves and to finance and build the supply system.

This concept does not appear very complicated to me. I wonder, however, how many of us have thought it through and expressed it in such terms.

In summary, we use minerals to provide some of the properties we need to perform the functions we require of materials. The sources of such properties are multiple. The availability of resources is therefore enormous, but we need massive efforts to turn resources into reserves through exploration, research, and development. And finally, we need extensive facilities (capital) to turn reserves into a supply system.

2

Eugene N. Cameron

The Contribution of the United States to National and World Mineral Supplies

It may seem odd that a book devoted to the future mineral position of the United States, having begun with a discussion of present use of minerals, should retrogress to a review of the role of the United States as a supplier of minerals in the past. There are, however, cogent reasons for such a review. The first reason is that the mineral position of the United States at any given time is a direct consequence of its history of mineral production and consumption. The second reason, really an extension of the first, is that the future mineral position of the United States will lie within limits that have already been partly determined by the developments of the past, and these must be kept in mind in appraising its future mineral position. The third reason is that there is current a misconception as to the role of the United States in supplying its own needs for minerals and in supplying minerals to other countries of the world. The United States is often depicted as a nation that has been rapidly devouring the world's mineral resources. This is a rather serious accusation. For all these reasons we must examine the mineral history of the United States, and we must appraise the role of the United States not only as a supplier of minerals to other countries of the world, but also as a consumer of minerals produced both within and without its borders.

The mineral history of the United States is primarily the record of its mineral production and its mineral consumption. If the two had always been the same, the record could be very simply stated, and little discussion of it would be required. But the United States has never been self-sufficient in minerals. Owing to this circumstance, the country has been involved in the international exchange of raw materials during many decades. Throughout the twentieth century it has operated in the context of world patterns of mineral production and mineral consumption, and these patterns must be considered in forecasting the future mineral position of this country.

When the twentieth century began, the United States, relative to the rest of the world, was already a major producer and consumer of mineral commodities (see fig. 2.1). During the period 1900–1904, total mineral production, measured in 1967 constant dollars, amounted to nearly 3.3 billion dollars annually. Already, however, deficiencies in United States mineral resources had appeared, and materials amounting in value to about 10 percent of United States mineral production were being imported from various countries. Imports, however, were more than balanced in value by mineral exports.

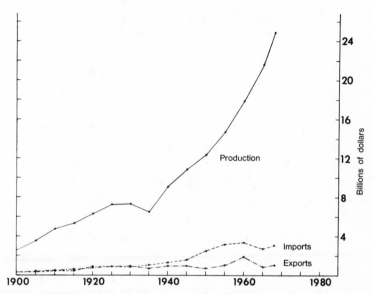

Figure 2.1. United States production, exports, and imports of all mineral commodities, 1900–1969, in constant 1967 dollars. Each point on a curve gives the average for the five-year period ending with the year for which the point is plotted. Data from Vivian E. Spencer, *Raw Materials in the United States Economy, 1900–1969*, U.S. Bureau of the Census and U.S. Bureau of Mines, Working Paper 35 (Washington, D.C.:U.S. Government Printing Office, 1972).

As the 1900s, 1910s, and 1920s passed by, United States mineral production grew apace, responding to the burgeoning of industry in the nation and to economic development abroad. Between 1900 and 1929 the value of annual mineral production, again measured in 1967 dollars, more than doubled. The value of the country's annual mineral imports tripled, but so did the value of its mineral exports. A balance between mineral imports and mineral exports was thus maintained. Measured in these general terms, therefore, there was little change between 1900 and 1929 in the degree to which the United States was able to supply the demands of its expanding industries. If its position did not improve, at least on balance it did not grow worse.

The accusation that the United States achieved industrial power at the expense of the mineral resources of the rest of the world is therefore patently untrue for the period 1900–1929. During this period the United States produced 89 percent of all the minerals it consumed, and traded on an equal basis for the rest. It imported large amounts of manganese, chromite, tungsten, nickel, asbestos, and certain other minerals, but exported large amounts of other minerals, among them copper, lead, zinc, petroleum, phosphate, and sulfur.

The Great Depression of the 1930s ended this epoch of the mineral history of the United States. Since 1939 United States mineral production has continued to increase, more than doubling in constant dollar value since 1945. The balance between imports and exports, however, has not been maintained. The United States is an important supplier of coal, molybdenum, phosphate, boron minerals, and sodium minerals to the world, but by 1969 the value of its mineral exports had decreased to only about 8 percent of the value of its greatly increased total mineral production. Meanwhile, since 1945 the value of its mineral imports has ranged from 10 to 25 percent of the value of its mineral production, and in recent years has been more than three times the value of its mineral exports. Measured in dollar value, about 85 percent of its annual mineral needs are still coming from its own mineral deposits, so that the accusation that the United States has been devouring the world's minerals is again refuted by the record. The country's needs are now far greater, however, than thirty years ago; hence it requires greater amounts of minerals annually from foreign sources, and its dependence on those sources is increasing. Furthermore, its surplus production of certain minerals is no longer large enough to enable it to trade for mineral imports on an even basis. The significance of international trade in minerals to the United States has therefore changed since World War II.

It is apparent that the overall mineral position of the United States has deteriorated since 1930, and especially since the end of World War II in 1945. The nature of the deterioration and the present mineral position are important, and we must now examine them. This is no simple matter. We deal here with the self-sufficiency of the United States in mineral raw materials, and this is not easy to evaluate. Self-sufficiency in a mineral material for a given period is customarily calculated as the ratio of production from domestic deposits to domestic consumption, the ratio being expressed as a percentage. Self-sufficiency

figures calculated in this manner have certain draw-
backs, which can be illustrated by the case of iron ore.
United States reserves of iron ore are currently being
given, very conservatively, as about ten billion tons,
equal to at least a hundred years' supply at current
rates of consumption. Yet in 1971 the United States
imported about a third of the iron ore it consumed,
for a self-sufficiency rating of only 66 percent. In
this case, self-sufficiency is obviously not a measure
of what United States could produce annually if
it had to. The current self-sufficiency percentage
is actually the result of a delicate balance of economic,
technological, and political factors that determine the
iron ore supply structure of the American iron and
steel industry. A shift in one or more of these factors
can markedly increase or decrease the degree of United
States self-sufficiency over relatively short periods.

This is one problem. A second problem with iron
ore, and with certain other mineral materials, is that
in any one year nearly 30 percent of the total output of
iron and steel is from recycled scrap. "Consumption"
is thus used only in the narrow sense of consumption
of newly mined materials. Actual United States
self-sufficiency in iron and steel is therefore appreciably greater than suggested simply by matching
annual production of iron ore against annual consumption of iron ore. A final problem is that the use
of one mineral is integrated with the use of others.
Shortages of one mineral may drastically affect the
scale of use of others. Self-sufficiency with respect to a

mineral can therefore be evaluated fully only in the context of its use.

One may well ask: Do self-sufficiency percentages obtained by matching production against consumption have any real value? The answer has to be "yes." They do give a measure of the ability of the United States to fulfill its mineral requirements from domestic sources that are competitive in international mineral markets. They are a rather good measure of this ability over any extended period, and they are the only measure that is at all satisfactory.

What is the record in the period since World War II? Figure 2.2 is one way of answering this question. The chart compares, for forty-one mineral commodities, United States self-sufficiency in 1947–48 with that in 1970–71. Let us consider them in groups. Taking the first eight minerals, the United States was self-sufficient or better in all of these in 1947–48; in 1970–71 ratios of production to consumption were even higher. In the next twelve minerals, the country was deficient in 1947–48. Improvements in self-sufficiency are shown in 1970–71, but changes, except for industrial diamonds, are not marked, and in most cases little has been done to correct the deficiencies of 1947–48. For the next group of six minerals, self-sufficiency has declined, but not seriously. For the following eight minerals, plus rutile, however, marked declines in self-sufficiency are shown. Particularly significant are growing deficiencies in domestic supplies of petroleum, iron ore, fluorspar, bauxite, and zinc. The

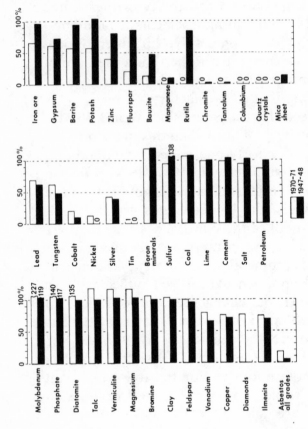

Figure 2.2. United States self-sufficiency in forty-one mineral commodities, 1970–71 compared with 1947–48. Self-sufficiency is expressed in percent, and for each of the two periods is equal to

$$\frac{\text{Production from domestic mines}}{\text{Consumption of newly mined material}} \times 100.$$

Data from U.S. Bureau of Mines, *Minerals Yearbook* and *Commodity Data Summaries* (Washington, D.C.: U.S. Government Printing Office, annual volumes).

other six minerals on the chart the United States did not produce in important amounts in 1947–48; there has been no improvement since.

The changes in self-sufficiency are therefore very mixed. Three things stand out, however. First, even though the United States still produces, in dollar value, about 85 percent of its total mineral needs, there are now serious deficiencies in more than half the minerals on the chart. Most of the deficiencies are in the metals. These have been a supply problem for many decades. In 1952 the President's Materials Policy Commission pointed out that the United States was then self-sufficient only in two metals, molybdenum and magnesium. The statement is also true in 1972, but

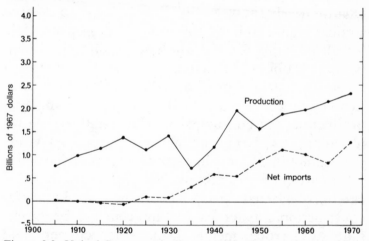

Figure 2.3. United States production and net imports (imports minus exports) of metals, 1900–1969, in 1967 constant dollars. Each point on a curve gives the average for the five-year period ending with the year for which the point is plotted. Data from Spencer, *Raw Materials in the United States Economy*.

with the difference that, as shown in Figure 2.3, net imports of metals have increased. In 1969 approximately one-third of the country's net supply of primary metals and metallic ores was obtained from sources abroad. Finally, on balance, the individual changes shown in Figure 2.2 between 1947–48 and 1970–71 sum up to the growing gap between United States mineral exports and United States mineral imports that is expressed in Figure 2.1.

What is this country's mineral position relative to the rest of the world? The record here is very plain. Just as the Great Depression was a turning point in the mineral history of the United States, World War II was a turning point in the mineral history of the world. Qualitatively this has been evident for some time. Quantitatively, however, it is not so easy to demonstrate, because comprehensive and homogeneous data for all minerals for the whole world are nearly impossible to obtain. To simplify this task, for some years I have been using a group of eighteen selected minerals as a key to world and United States trends in mineral production and consumption. The results are shown in Figure 2.4. Note that mineral production is expressed not in dollars but in tons. Changes in currency values are therefore not involved. The minerals are a carefully selected group. They include the major constructional materials iron and cement; the important nonferrous metals lead, zinc, copper, and aluminum (bauxite); the ferro-alloy metals manganese, chromium, molybdenum, vanadium, and nickel;

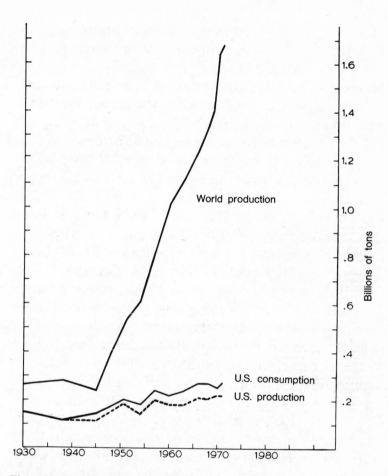

Figure 2.4. World production and United States production and consumption, 1930–71, of eighteen minerals (iron ore, bauxite, copper, lead, zinc, tungsten, chromium, nickel, molybdenum, manganese, tin, vanadium, fluorspar, phosphate, cement, gypsum, potash, and sulfur), in metric tons. Data compiled at the University of Wisconsin by Kenneth D. Markart and E. N. Cameron, from U.S. Bureau of Mines, *Minerals Yearbook* and *Commodity Data Summaries*.

and the major fertilizer minerals phosphate and potash. Some of the minerals are produced in large amounts in the United States, others are imported largely or entirely from abroad. All are key minerals in United States and world industry. Enlarging the list would alter the positions of the curves but would not change their relative positions or their trends. Neither would omission of iron ore and cement, the two largest contributors to the total tonnages indicated on the chart.

The chart shows that since 1945 United States production of minerals in this group as a whole has risen rather steadily, and production in 1971 was approximately double that in 1945. Consumption of this group of minerals, however, has risen even more rapidly, and the widening gap between production and consumption is another measure of the increasing dependence of the United States on foreign sources of minerals. For the world, however, the picture is quite different. Since 1945, world annual production, and this is essentially equal to world annual consumption, has risen at a very rapid rate, doubling at the rate of once every seven to eight years. In 1945 United States production was 46 percent of the world supply of the eighteen minerals. In 1971 United States production was only about 13 percent. The United States, therefore, is no longer a dominant factor in world production of this group of eighteen important minerals. The marked increase in world consumption reflects the increase in world population, but it likewise reflects an increase in per capita consumption, from about .11 tons in 1945 to about .44 tons in 1971.

From the standpoint of the influence of the United States on international trade in minerals, the United States role in world mineral consumption is important. In examining this, again we can take annual world consumption as approximately equal to world mineral production. On that basis, in 1945 the United States consumed nearly 60 percent of the world total for the eighteen minerals. Since 1945, however, the United States percentage of world consumption has declined almost steadily. In 1971 the figure was 16 percent. The United States began the post-World War II period as by far the largest market for the mineral raw materials of the world. It is still an important market, but its position of dominance is lost.

The chart we have just examined involves only the mineral raw materials. What about sources of energy? Three charts are pertinent. One (fig. 2.5) is a chart showing changes in amounts and sources of energy consumed annually by the United States during the postwar period. The chart records two things: (1) the drastic increase in annual United States energy production and consumption, from 7 quadrillion B.T.U. in 1900 to 69 quadrillion B.T.U. in 1971, and (2) the increasing importance of petroleum and natural gas in energy production in the United States during the period. In 1970 petroleum alone accounted for 44 percent of United States energy production.

Figure 2.6 shows United States production and consumption of coal, together with world production.

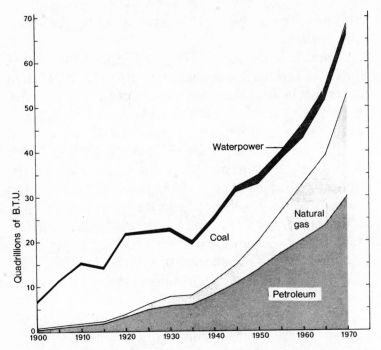

Figure 2.5. United States sources of energy, 1900–1970. The contribution of nuclear energy is too small to be shown. Data from U.S. Bureau of Mines, *Minerals Yearbook*, and from U.S. Department of the Interior, *First Annual Report of the Secretary of the Interior under the Mining and Minerals Policy Act of 1970* (Washington, D.C.: U.S. Government Printing Office, 1972), pt. 1, p. 19.

United States production of coal has exceeded consumption by about 10 percent during the postwar period. The United States is no longer, however, a dominant figure in world production; it has, in fact, been replaced by Russia as the world's largest producer.

Figure 2.6. World production and United States production and consumption of coal, 1931–70, in short tons. Data from U.S. Bureau of Mines, *Minerals Yearbook* and *Commodity Data Summaries*.

In natural gas the United States has been essentially self-sufficient since 1900, but, as shown in Figure 2.7, that is not the case for crude petroleum. In 1945 oil wells in the United States yielded about 62 percent of total world production, and the country was nearly self-sufficient in petroleum. By 1971 United States production had doubled, but it was producing only about 19 percent of the world total. Here, as in mineral raw materials and coal, the United States has lost its position as the world's principal producer. It

may actually have passed the peak of its petroleum
production in 1971. The chart also records the widening gap between United States production and consumption. Finally, implicit in the chart is that most of

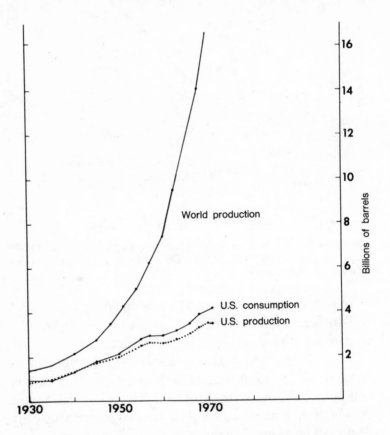

Figure 2.7. World production and United States production and consumption of crude petroleum, 1930–71. Data from U.S. Bureau of Mines,
Minerals Yearbook and *Commodity Data Summaries*.

the world's petroleum production is being consumed by other nations than the United States.

In summary, the following have been the principal developments in United States mineral history and world mineral history since 1939:

1. United States mineral production has greatly increased but has not kept pace with consumption. Self-sufficiency in minerals has declined, both overall and in numbers of minerals involved.

2. Since 1945, world mineral production has increased far more rapidly than United States production. The relative importance of the United States as a supplier of world minerals, raw and manufactured, has rapidly decreased.

3. World mineral consumption has increased far more rapidly than United States consumption. The United States is no longer the world's principal market for mineral raw materials, and it faces increasing competition for the world's mineral supplies.

At the present time efforts are being made, through the National Materials Policy Commission, to develop a comprehensive mineral policy for the United States. In developing this policy, the history that we have reviewed must be taken into account. The history since World War II deserves particular attention. Under the stimulus of increased demands for minerals from United States industry and the American public, this was a period of unparalleled activity in mineral exploration in the United States. Huge amounts of money and large-scale, intensive efforts were brought to bear on the problems of mineral exploration. Geological, geochemical, and geophysical techniques of exploration that were unavailable or poorly developed prior to World War II were devised or perfected and put to use. Furthermore, restrictions on land available for prospecting and exploration were less than at present, and limitations on mining arising out of environmental concerns imposed fewer constraints on mineral development during the period than is likely in the future.

Advances were not confined to methods and scale of mineral exploration. Parallel with this activity enormous strides were made in mining methods and in methods of processing ores and extracting valuable minerals from them. These advances contributed heavily to the physical volume of material flowing from American mines, and they broadened the range of earth materials economically available to man.

What, then, has been the result? Actually, the record has been one of spectacular success, measured in terms of the size and significance of new mineral deposits discovered and new kinds and grades of materials brought into the useful range. The doubling of United States annual production of minerals during the period is the record of this success. It is all the more remarkable when we recall that when World War II ended in 1945, United States reserves of important minerals such as iron ore, lead, zinc, copper, and tungsten had been seriously depleted owing to the intensive mining that was necessary during World War II.

Yet despite this success, the United States has not been able to correct its mineral deficiencies and has fallen steadily behind in its ability to supply its needs. This is a cardinal fact that must be faced in present efforts to frame a national mineral policy. If the United States has had a mineral policy in the past, which is certainly a debatable question, it has consisted of encouraging production from domestic mines and relying on sources abroad for what the country could not produce itself. The historical record indicates, however, that a more positive approach to the problem of mineral supply will be needed if the United States is to arrest the deterioration in its mineral position and is to maintain a strong mineral position during the remainder of the century. A comprehensive policy is required; isolated actions and the piecemeal policy-making of the past are unlikely to be successful. It is necessary, first, to define the role of minerals in the future economy and culture of the United States and, second, to devise a comprehensive, positive policy that will ensure availability of necessary mineral supplies in adequate amounts. This is a formidable undertaking, but it will be even more formidable if postponed. Furthermore, if it is delayed too long, we may well find that certain options open to the country now have been foreclosed.

3 *John D. Morgan, Jr.*

Future Use of Minerals: The Question of "Demand"

The earth and the sun are the fundamental sources of all wealth. Demand is limited absolutely by what can be obtained from the rocks, soils, waters, and atmosphere of the earth, plus earth's share of solar energy. Rising standards of living in the technologically advanced nations, coupled with the desires of the burgeoning populations of the less developed nations, focus increased attention on demand for minerals, both now and into the foreseeable future.

Man's technological progress has been so dependent upon minerals that major historical periods are named the Stone Age, the Bronze Age, the Iron Age, the Steel Age, the Age of Light Metals, and the Nuclear Age, marking the new mineral materials added to those that had been in use from the dim past of

Note: Opinions are those of the author and do not necessarily reflect official views.

prehistory. However, in terms of tonnages in use and annual new tonnages required, we are still in the Stone Age, because about half of United States annual demand for new mineral supplies involves sand, gravel, and stone.

The original inhabitants of the three and a half million square miles that today is the United States used a number of minerals, including clay, natural asphalts, rock oil, red and yellow iron oxides, granite, flint, and turquoise. But their total annual demand for all those minerals would have totaled only a few tens of tons. By 1950 United States annual demand for new mineral supplies including fuels had reached two billion tons; by 1971 it had doubled to four billion tons; and projections of demand point to annual United States demand of eleven billion tons by the year 2000.

It is important to consider demand not just in overall quantitative terms but also in value terms. Very different values can be ascribed to United States mineral demand, depending upon the stage of processing. Figure 3.1 shows that 1971 United States mineral demand was estimated to be $34 billion in the form of such raw materials as crude oil, coal, iron ore, bauxite, sand, gravel, and stone. This $34 billion demand was supplied by $30 billion of domestic productions plus $4 billion of imports. At the next stage of processing, however, 1971 United States mineral demand was estimated to be more than $156 billion in the form of processed materials of mineral origin

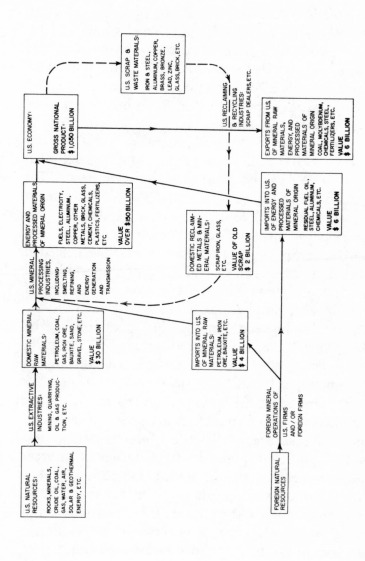

Figure 3.1. The role of minerals in the United States economy (estimated values for 1971)

and energy. Processed materials of mineral origin include such items as steel, nonferrous metals, concrete, glass, brick, tile, fertilizers, chemicals, and plastics. Energy includes solid and liquid fuels and electricity. This $156 billion demand was supplied by domestic production valued at $150 billion plus imports valued at $6 billion.

Demand at each stage of processing must be considered because such demand can often be supplied from different sources. For example: iron needed for domestic blast furnace feed can be domestic or foreign ores, concentrates, or pellets; iron needed for domestic electric furnace feed can be domestic or foreign pig iron, scrap, or pre-reduced pellets; steel needed by manufacturers can be domestic or imported steel and steel products. Further, the advanced stages of mineral processing are specifically included in the National Mining and Minerals Policy established by Public Law 91–631, approved December 31, 1970, which refers to the "domestic mining, minerals, metal and mineral reclamation industries."

Some United States demand is created at even more advanced stages of manufacturing, as for example when manufactured articles of foreign origin that contain significant quantities of materials of mineral origin are desired and imported. While it becomes increasingly difficult to identify the values attributable to minerals per se as the stages of manufacturing become more advanced, it is obvious that minerals and agricultural products together are the fundamental materials that make possible the United States Gross National Product, which was valued at $1,050 billion in 1971.

National security considerations must be broadly viewed. Politico-economic factors must be weighed along with purely military ones. The United States should not allow itself to become too dependent upon potentially unreliable sources of supplies. Remedial measures could include maintenance of some assured level of domestic production, stand-by facilities, stockpiles, development of substitutes, and/or stand-by controls. National security considerations, broadly viewed, may also place special politico-economic restrictions on mineral demand; for example, favoring demands for minerals produced in some sources while discouraging demands upon other sources. Balance-of-payments considerations can affect demand for minerals which contribute either positively or negatively thereto. Politico-economic considerations affect demand for monetary metals.

The needs of specialized government programs (such as military, aerospace, and nuclear) must be assessed. Demand for special-property minerals for such programs can skyrocket, as evidenced in the past by beryllium, hafnium, titanium, tungsten, uranium, and zirconium. When such programs are cut back, demand, like a spent skyrocket, can fall rapidly, Nevertheless, it is essential that government and industry make continuing efforts to identify potential increased demands for special property minerals so that appropriate supply expansion programs can be launched in a timely and orderly manner.

Real demand for minerals must bear some relationship to actual ability to pay for them and the ability of the earth to supply them. Further, real demand and real supply are by no means likely to be in phase. Indeed, except in defense emergencies, over past decades and continuing to the present, short-term availabilities have been such that producers of most minerals have made major efforts in research, advertising, and marketing to increase demand in old and in new applications. In some instances, most often in defense emergencies, national demands for defense programs were quantified by governments, and major supply-expansion programs were initiated to increase supplies to meet new levels of anticipated demand. In the United States the most recent such programs were those involving accelerated tax amortization, market and price guarantees, long-term contracts, and exploration assistance during the Korean War period. Minerals are not ordinarily produced unless there is a foreseeable market for them, nor are major exploration efforts made unless there is a reasonable possibility of marketing the minerals sought. Much of the world is yet to be systematically investigated in detail as to its surface and near-surface geology. Our deepest mines have penetrated only about two miles of the earth's crust and our deepest wells less than six miles, while our seagoing dredges operate in only a few hundred feet of water.

Demand for specific minerals may vary signifi-

cantly from year to year in response to economic
conditions, changes in laws and regulations, or consumer preferences. For example, in the United States
the automotive industry generates about one-fifth of
the total annual demand for steel and proportionate
quantities of other minerals. In 1971 there were about
110 million motor vehicles of all types on the nation's
roadways. While the specifications for vehicles vary
widely, the average standard American auto is made
up approximately as follows:

2,775 lbs.	iron and steel (frame, engine, and body)
100 lbs.	aluminum (components and some engines)
50 lbs.	copper (radiators and electrical)
25 lbs.	lead (battery)
50 lbs.	zinc (castings, galvanizing, and tires)
100 lbs.	glass (windshield and windows)
250 lbs.	rubber and plastics (tires and fittings)
150 lbs.	miscellaneous (fabric, insulation, etc.)
3,500 lbs.	Total

Almost all of the above are derived from minerals,
including the plastic and synthetic rubber. Consequently, American 1971 production of 8.6 million
autos, plus 2.1 million new commercial vehicles, generated mineral demands substantially greater than
10.7 million times the above unit quantities. The demand for processed materials of mineral origin substantially exceeds the drive-away weight of the vehicle
because there are unavoidable losses in manufacturing
operations.

The construction industry is another generator of
major demands for minerals. When construction of

houses, hotels, factories, roads, airports, power plants, harbors, etc., is accelerating, the effect is felt throughout the mineral industry. Through fiscal, home financing, public works, and urban renewal programs the United States government can substantially influence the pace of construction. Changes in building codes to accommodate newer and composite materials and modular construction methods can have major impact on demand for minerals. In building, the common denominator for judging materials is "dollars per square foot of floor area or cubic foot of space."

Attempts to assess demand must take cognizance of rapidly changing patterns in usage. Modern industrialized civilizations have the ability to solve many problems by alternate combinations of resources. For example, Table 3.1 illustrates the wide variety of materials that are used in building ordinary houses and in heating them. All of these materials are of mineral origin. The choice of a given material for a particular use is governed partly by price, partly by specific properties and specifications, and partly by assured availability.

Alternate methods and procedures must also be considered in assessing demand. For example, food may be preserved by packing in "tin" cans, which are actually tin-coated steel. Food may also be preserved by packing in aluminum cans or lacquer-lined steel cans, or in glass jars or bottles. Beverages are packaged in all-aluminum or steel-aluminum cans, or in glass or plastic bottles. Many foods may be preserved by

Table 3.1. The Wide Variety of Mineral Materials Used in Homebuilding

Building materials	Insulating materials	Heating and cooling materials	Plumbing materials
Sandstone	Asbestos	Burning coal (heat only)	Plastic pipe
Limestone	Rock wool	Burning oil	Soft copper tubing
Marble	Fiberglass	Burning natural gas	Copper pipe
Granite	Plaster (gypsum)	Burning manufactured gas	Brass pipe
Concrete	Plaster board (gypsum board)	Electricity, in turn derived	Black iron pipe
Common brick	Vermiculite	from:	Galvanized iron pipe (zinc-coated)
Tile	Fly-ash block	Coal	Cast iron pipe
Concrete block	Aluminum foil	Oil	Lead pipe (waste lines only)
Pebbles set in cement	Aluminum shade screening	Natural gas	Lead sheet (showers, etc.)
Asbestos-cement siding	Bronze shade screening	Manufactured gas	Glass pipe
Slate granule siding	Tar paper	Hydropower	Zinc die-castings
Painted steel	Double pane glass	Geothermal power	Chromium-plated brass
Galvanized steel (zinc-coated)	Plastic sheet	Nuclear power	Nickel-plated brass
Oxidizing steel	Foamed plastic	Solar Energy (rarely by itself)	Chromium-plated iron
Stainless steel	Magnesite	Etc.	Stainless steel
(chromium-nickel)	Foamed glass		Vitreous china
Plastic-coated steel			Porcelain ware
Glass			Porcelain-coated steel
Aluminum			Cement-asbestos pipe
Plastic-coated aluminum			Concrete pipe
Solid plastic			Clay tile pipe
Fiberglass plastics			Plastic sheet
Asphalt shingles			Ceramic tile
Asphalt roofing			Plastic parts and fittings
Etc.			Etc.

freezing, wrapped in plastic, paper, or aluminum foil. Some foods are still preserved by the older methods of smoking, salting, baking, or drying. Demand in a variety of industries can be achieved by a variety of means.

Export demand is another item that must be considered. The United States in the past has been an important exporter of several mineral raw materials, notably coal, molybdenum, phosphate, sulfur, and vanadium, and many processed materials of mineral origin, notably metals and chemicals.

Not all demands for minerals are increasing. Some indeed may well decline. More questions are being raised as to whether there may not be too much of some items in circulation in certain applications, or perhaps misapplications. In recent years radioactive materials, sulfur oxides, carbon monoxide, nitrogen oxides, hydrocarbons, photochemical oxidants, various particulates in the air, mercury, lead, cadmium, beryllium, phosphorus, asbestos, and many chemicals are among materials subject to restrictions which should curtail demand. Conversely, pressures to improve the environment may stimulate demand for other materials. There is also increasing concern about displaced earth and rock in connection with slides, leaching, and erosion. Any overall calculations of future mineral demand must include realistic balancing of demands for cleaner air, cleaner water, restoration of mined lands, extinguishment of mine fires, rehabilitation of tailing piles, maintenance of wilderness

areas, parks, and recreation sites, and disposition of
large stocks of materials accumulated pursuant to
environmental regulations—as in the case of sulfur.

Steel production is a major index of modern industrialization. Iron, the predecessor of steel and the
most important element in steel, has been in use by
man for over four thousand years. Iron is one of the
more abundant elements, accounting for over 5 percent of the earth's crust. Steel possesses a reasonably
high melting point, and its alloys are extremely tough
and resistant to most physical forces. While steel
manufacturing often involves mammoth installations,
its technology is relatively simple, and steel is a relatively cheap material. Figure 3.2 diagrams the growth
of United States and world steel production over the
past two decades. In 1950 the United States produced
about one-half of the world's steel. Thus, two decades
ago, United States demand was about one-half of
total world demand for all raw materials. Also, two
decades ago the United States produced about one-
half of all the world's hard manufactured articles.
Consequently, two decades ago the United States was
a major factor both in world markets for mineral raw
materials and for manufactured goods. Today, producing only about one-fifth of the world's steel, despite
a quantitative increase over 1950, the United States
has far less leverage in world mineral markets when
it comes to spot purchasing or developing new sources
abroad. Also the United States is encountering far
greater competition when it comes to selling manu-

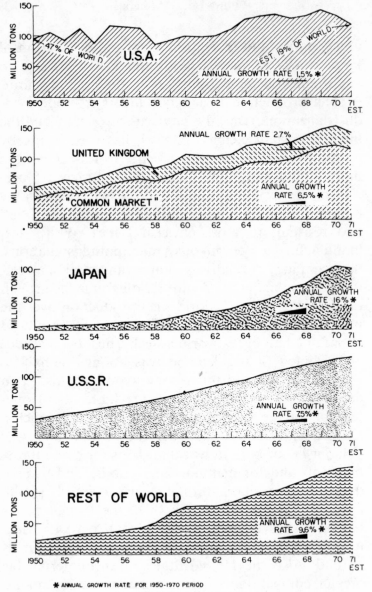

Figure 3.2. World steel production, 1950–71

factured articles. Indeed, competition is so severe that
in 1971 18.3 million tons of imported steel supplied
about 15 percent of United States domestic demand
at the same time that 2.7 million imported motor vehicles, each equivalent to about one ton of steel, also
entered the United States. As shown by Figure 3.3,
increased demand for iron in the United States over
the past two decades has been met largely by a com-
bination of increased imports of iron ores and pellets
and steel products.

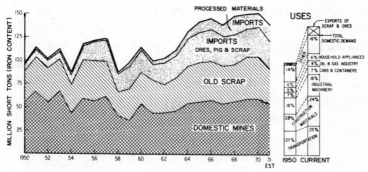

Figure 3.3. United States supplies and uses of iron, 1950–71

Another major group of metals also fundamental
to all industrialized economies consists of the older
nonferrous metals copper, lead, and zinc, plus the
newer metal, aluminum. Copper has been in use for
over five thousand years. Bronze, about 85 percent
copper and 15 percent tin, gave its name to an early
period. Strong corrosion-resistant bronze alloys of
today are little different from those made in Phoenicia

four thousand years ago. Lead has been in use for over
two thousand years, and many current techniques were
widely employed in ancient Rome. Zinc has been in
use in the Western world for over five hundred years
and in China long before. The manufacture and use
of white zinc oxide was described by Marco Polo,
while the composition of old Chinese brass, about 70
percent copper and 30 percent zinc, is the current
formulation for modern brass alloys. Copper, lead,
and zinc, however, are relatively rare elements of the
earth's crust, all three together totaling less than a
small fraction of one percent. In contrast, aluminum,
first produced commercially toward the close of the
nineteenth century, accounts for over 8 percent of the
earth's crust. Aluminum is the third most abundant
element in the earth's crust, being exceeded only by
silicon (28 percent) and oxygen (47 percent). Figure
3.4 shows United States demand for and supplies of
aluminum, copper, lead, and zinc over the past two
decades. Clearly evident is the more rapid increase in
the use of aluminum, reflecting its relative strength-
weight, workability, and corrosion-resistant properties, as well as its comparatively reasonable price.
Because aluminum is only about one-third the density
of steel, its base price of about three times that of
steel permits competition when other properties are
comparable.

In the past two decades, as shown by Figure 3.5,
there has been a very rapid rise in the demand for plastics. Most plastics are mineral-based products, in that

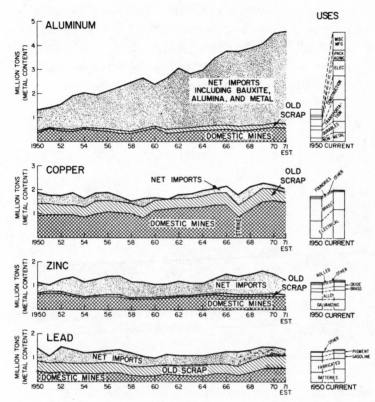

Figure 3.4. United States supplies and uses of aluminum, copper, zinc, and lead, 1950–71

they are composed largely of carbon and hydrogen derived primarily from petroleum and natural gas. Fiberglass reinforcing of plastics creates another demand for minerals. The annual tonnage of plastics used in the United States now exceeds the combined tonnages of aluminum, copper, lead, and zinc. Inasmuch as most

plastics are much less dense than metals, they are used
where volume or weight considerations are significant.
Because of the differences in density it is possible for
a plastic, sold at significantly higher per pound prices,
to substitute for a lower priced metal, and on a volume

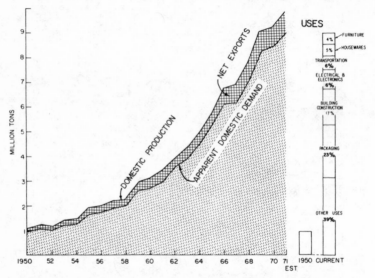

Figure 3.5. United States supplies and uses of plastics, 1950–71

basis United States annual demand for plastics is equal
to about half of demand for steel.

Major demands for nonmetallic minerals are generated by the construction and the chemical industries.
Our bountiful agriculture is in large measure attributable to abundant use of mineral fertilizers. Demand
for refractories and abrasives creates additional demand for nonmetallic minerals. Demand for non-

metallics currently accounts for half of all tonnage of United States mineral demand. The wide use of silicate materials reflects the fact that silicon accounts for about one-fourth and oxygen about one-half of the

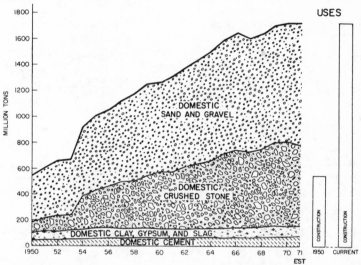

Figure 3.6. United States supplies and uses of major nonmetallic construction materials, 1950–71

earth's crust. Concrete, consisting of cement, sand, and gravel, is a major construction material which directly competes with metals, as well as requiring use of metal reinforcing in connection therewith. Figure 3.6 shows the threefold rise in United States use of major nonmetallic construction materials over the past two decades, while Figure 3.7 covers the major fertilizer elements. Technologic change and improvement has had a major impact on the use of some nonmetallic

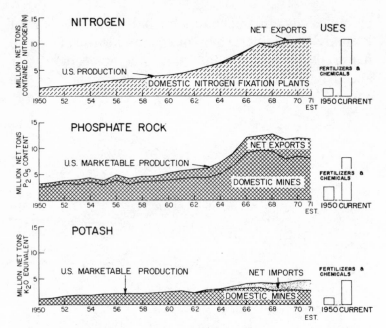

Figure 3.7. United States supplies and uses of major fertilizer ingredients, 1950–71

minerals. Synthetic substitutes for domestic materials in short supply have been developed to take the place of sheet mica, natural diamonds, natural abrasives, natural refractories, and natural piezo-electric crystals.

The amount of energy now used in the United States is equivalent to the work of three hundred persons laboring round the clock for each one of our more than 208 million citizens (70×10^{15} B.T.U. converted at 0.05 hp. per worker). As shown by Figure 3.8, United States demand for energy has doubled over the past two decades. This demand for energy has resulted in

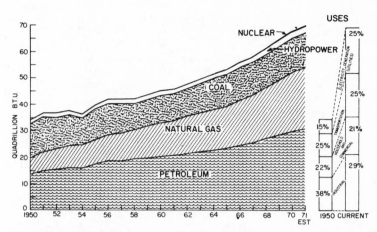

Figure 3.8. United States demand for major energy sources, 1950–71

relatively steady demand for coal and in significantly increased demand for natural gas and petroleum, as illustrated in Figure 3.9. Figure 3.10, uranium, illustrates the major expansion that was sponsored by the A.E.C. in the past two decades. In energy, too, the possibilities of substitution are significant, as power plants can be designed to burn oil, coal, or natural gas interchangeably; liquid fuels can be made from agricultural products as well as coal, gas, and petroleum; and gaseous fuels can be made from solid or liquid fuels. Carbon and hydrogen can be linked in many ways to form solids, liquids, or gases. Demand for energy could be substantially modified by presently known conservation possibilities. Many homes and public buildings are too warm in winter and too cold in summer for optimum health. Greater use of insu-

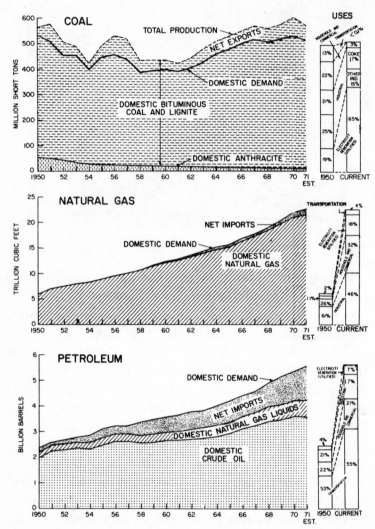

Figure 3.9. United States supplies and uses of coal, natural gas, and petroleum, 1950–71

lation, including double windows, storm doors, and shade screening devices will reduce heat loss in winter and heat gain in summer. Smaller vehicles and more use of mass transportation could reduce demand for

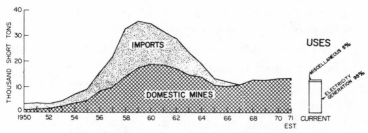

Figure 3.10. United States supplies and uses of uranium (U_3O_8), 1950–71

liquid fuels. The common denominator for judging energy demands and energy substitution possibilities is "cents per million B.T.U.," modified by consumer preferences based on convenience and local availability.

In the grand scheme of nature it would be more appropriate to consider "demand for elements" rather than "demand for minerals" because the living creatures of the earth are inextricably intertwined in the dynamic ecosystem in which the ninety-two naturally-occurring elements are constantly combining, separating, and recombining in response to fundamental forces. Hematite, magnetite, limonite, siderite, etc., are important minerals, but iron is the more important common elemental constituent of interest. Anthracite, bituminous coal, lignite, peat, and so on are important minerals, but carbon is the most important common elemental constituent of interest. Petroleum and natural gas are important mineral fuels, but carbon and hydrogen are the more important common elemental constituents of interest. Nevertheless, it is still very necessary to consider demand for and supplies of individual minerals because different minerals occur in different ways, different geological, geophysical, and geochemical methods are required to discover them, different beneficiation methods are required to separate and concentrate them, and different chemical and metallurgical processes are required to convert them into more useful forms. Further, many commercial transactions still deal largely with minerals per se. For example, columbite, tantalite, ilmenite, rutile, zircon, potash, barite, bauxite, corundum, feldspar, fluorspar, kyanite, magnesite, quartz, talc, vermiculite, etc., are still common items of trade. In cases where a mineral is used as a consequence of chemical and/or physical properties deriving from its unique composition, it is essential to maintain the mineral identification. In contrast, where the main interest in a mineral derives from its content of a particular element which is used in elemental form, it is essential to refer to content, as, for example, percentage contained of copper, lead, zinc, iron, and so on, or ounces contained per ton, as in the case of precious metals. Consequently, the prevailing use patterns govern to a large extent whether assessment of demand in elemental or in mineral terms is most revealing; often assessment in both ways is required for complete understanding. For example: production and trade in salt per se must be considered because salt is found, mined, processed, and sold as salt. Salt is a major tonnage item, and United States annual demand therefor is almost fifty million tons. Yet it is also important to analyze demand for and supplies of sodium and chlorine, the two elements in common salt of particular interest to the chemical industry.

In assessing demand careful consideration must also be given to specialized regional situations. Some smelters, for example, can only treat certain types of ores, some refineries can only use certain grades of crude oil, some power plants can only handle certain types of fuel oil, and so on.

Also, in the case of large-bulk, low-cost items such as coal, many ores, sand, gravel, and stone, transportation costs may approach or exceed production costs. Consequently, regional demand analyses are essential because demand for some items in one region normally could not economically be supplied by excesses in supply from a distant, albeit also domestic, region.

Attempts to assess future needs are normally based on extrapolation of past usage patterns appropriately modified by foreseeable technological, economic, and social changes. In forecasting it is necessary to consider both overall national totals and identifiable components of total demand, so that the validity of projections can be checked by testing extrapolations of gross data against summations of extrapolations for major programs. Projections of demand must lie within the limits of reasonableness. If in recent years, for example, United States annual usage of a certain commodity was one hundred units, it would be unreasonable to forecast near-term future usage as two hundred units or fifty units without clear justification. Two hundred units could be justified if demonstrated major new uses are highly probable. Fifty units could be justified if significant available substitutes are about to be utilized. In the case of materials for which usage is closely tied to defense and aerospace programs, it would not be impossible for near-term future demand to jump from one hundred units to one thousand units or drop to ten units based upon firm programs.

Over the years the Bureau of Mines has made intensive efforts to assess past, present, and future demand. A recent such study by the Bureau of Mines was included in the *First Annual Report of the Secretary of the Interior under the Mining and Minerals Policy Act of 1970,*[1] and it is summarized herein by Figure 3.11. This figure shows United States primary mineral demand in the period 1950 through 1970, together with

1. U.S. Department of the Interior, *First Annual Report of the Secretary of the Interior under the Mining and Minerals Policy Act of 1970 (P.L. 91–631)* (Washington, D.C.: U.S. Government Printing Office, 1972). This report has been distributed to depository libraries. Sale copies may be ordered from: Superintendent of Documents, U.S. Government Printing Office, Washington, D.C. 20402. When ordering include title and specify: Pt. 1, 1972, I-1.96/3:972, S/N 2400–0739 ($1.25); Pt. 2, 1972 Appendices, I-1.96/3:972/pt. 2, S/N 2400–0738 ($3.25).

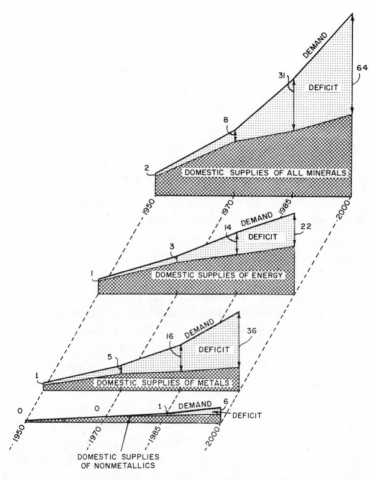

Figure 3.11. Developing deficits:United States primary mineral demand vs. United States primary mineral supplies (billions of dollars, with 1985 and 2000 at 1970 prices)

projections to 1985 and 2000. These demand projections are based on a 1985 United States population of two hundred fifty million with G.N.P. of $1,800 billion (in 1970 dollars), and a 2000 United States population of three hundred million with G.N.P. of $3,400 billion (in 1970 dollars). The supply projections are extrapolations of actual performance in the two decades 1950–70. Increasing future deficits appear evident from these projections, giving rise to greater concern as to our future mineral position. Item-by-item quantitative details are provided by Table 3.2 while conversion to values is provided by Table 3.3.

While there are major uncertainties inherent in any attempt to predict future United States demand for minerals, these difficulties are compounded when efforts are made to predict world-wide demand. The United States has only about 6 percent of the world's population and only about 6 percent of the world's land area. However, the United States now has a demand for about one-quarter of the current annual mineral production of the earth. Obviously, then, the other 94 percent of the people of the earth living on the other 94 percent of the earth's land area are currently using a smaller proportionate share of the earth's annual mineral production. Furthermore, the material standard of living in the United States is higher than in any other country of the world. Indeed, the best measure of a material standard of living is *not* annual new mineral supply but rather the per capita share of use of the *total materials in place*. For example, United States annual new copper "consumption" is reported to be about twenty pounds per person, but the term

Table 3.2. Comparison of United States Primary Demand with United States Primary Production: 1970, 1985, and 2000 (in thousands of units)

Commodity	Units	1970 actual primary demand	1970 actual primary production	1985 projected primary demand	1985 projected primary production[1]	2000 projected primary demand	2000 projected primary production[1]
Aluminum	ST	3,951	583	11,500	490	26,400	503
Antimony	ST	14.735	2.111	26.500	.370	48.000	0
Arsenic	ST	*	*	28.000	1.800	38.000	1.240
Barium	ST	787	478	1,000	460	1,380	419
Beryllium	ST	*	*	.820	.060	1.770	.082
Bismuth	lb.	2,237	*	3,000	740	4,000	708
Boron	ST	88	175	163	255	303	345
Bromine	lb.	342,000	350,000	414,000	457,000	500,000	600,000
Cadmium	lb.	9,108	3,555	16,600	4,400	30,800	4,510
Calcium	ST	89,618	89,660	147,000	102,000	241,000	118,000
Cesium	lb.	*	0	10.700	0	20.000	0
Chlorine	ST	9,754	9,755	20,300	14,300	42,300	19,800
Chromium	ST	462	0	700	0	1,120	0
Cobalt	lb.	16,190	*	20,000	0	24,700	0
Columbium	lb.	5,056	0	9,790	0	19,200	0
Copper	ST	1,572	1,720	2,900	1,910	5,400	2,380
Fluorine	ST	613	123	1,210	70	2,400	40
Gallium	kg.	*	*	.500	.490	.850	.702
Germanium	lb.	40	28	50	3	75	0
Gold	t. oz	6,147	1,743	9,200	1,270	14,300	1,000

Table 3.2. Comparison of United States Primary Demand with United States Primary Production: 1970, 1985, and 2000 (in thousands of units) (*continued*)

Commodity	Units	1970 actual primary demand	1970 actual primary production	1985 projected primary demand	1985 projected primary production[1]	2000 projected primary demand	2000 projected primary production[1]
Hafnium	ST	*	0	.038	0	.050	0
Indium	t. oz	*	*	650	300	800	394
Iodine	lb.	5,062	*	8,900	565	15,500	470
Iron	ST	84,000	59,000	113,000	51,000	153,000	50,000
Lead	ST	829	572	1,090	420	1,430	470
Lithium	ST	*	*	5.920	4.300	10.000	5.880
Magnesium, metal	ST	96	112	235	115	580	142
Magnesium, nonmetallic	ST	1,057	1,029	1,510	1,410	2,160	1,670
Manganese	ST	1,327	66	1,770	0	2,360	0
Mercury	fl	54	27	66	37	80	44
Molybdenum	lb.	49,104	111,352	96,500	140,000	188,000	184,000
Nickel	lb.	311,400	30,600	492,200	60,000	770,000	84,900
Nitrogen, compounds	ST	10,364	10,919	20,300	18,200	39,700	26,000
Nitrogen, gas & liquid	ST	5,310	5,310	10,500	8,500	20,900	12,700
Palladium	t. oz	539	10	780	28	1,060	39
Phosphorus	ST	3,630	5,236	6,590	8,500	12,000	11,700
Platinum	t. oz	407	7	634	0	1,000	0
Potassium	ST	3,908	2,265	6,850	3,660	12,000	4,640
Rare earths	ST	11.500	*	16.000	20.800	22.000	30.800
Rhenium	lb.	3,150	5,900	6,250	7,690	12,700	12,000

	Unit						
Rhodium	t. oz	36	0	55	1	85	2
Rubidium	lb.	*	0	1.200	0	2.200	0
Scandium	kg.	.015	0	.020	0	.025	.015
Selenium	lb.	1,105	975	1,340	916	1,620	988
Silicon	lb.	531	530	750	679	1,000	883
Silver	t. oz	73,100	45,000	124,000	39,000	210,000	40,000
Sodium	ST	20,337	19,668	37,000	27,000	67,200	36,000
Strontium	ST	*	0	25.600	0	34.200	0
Sulfur	LT	9,132	9,549	16,500	11,800	30,000	14,500
Tantalum	lb.	1,120	0	2,100	0	4,010	0
Tellurium	lb.	285	158	322	204	366	215
Thallium	lb.	*	*	8.000	3.910	9.000	4.600
Thorium	ST	*	*	.700	.049	1.500	.044
Tin	LT	53.027	*	70.000	.002	90.000	0
Titanium, metal	ST	24	0	65	0	168	0
Titanium, nonmetallic	ST	466	276	925	450	1,810	576
Tungsten	lb.	16,200	8,105	34,200	4,500	74,000	2,260
Vanadium	ST	7.066	5.594	14.700	10.000	31.000	13.000
Yttrium	ST	*	*	.250	.022	.420	.024
Zinc	ST	1,302	534	1,820	500	3,000	486
Zirconium, metal	ST	*	0	.010	0	.020	0
Zirconium, nonmetallic	ST	*	*	108	69	167	95
Asbestos	ST	734	125	1,340	198	2,430	273
Clays	ST	53,000	55,000	96,000	70,000	174,000	82,000
Corundum	ST	2.000	0	1.480	0	1.100	0
Diatomite	ST	444	598	942	963	2,000	1,250
Feldspar	LT	578	648	1,270	841	2,500	1,020
Garnet	ST	16	19	28	29	53	38

Table 3.2. Comparison of United States Primary Demand with United States Primary Production: 1970, 1985, and 2000 (in thousands of units) (continued)

Commodity	Units	1970 actual primary demand	1970 actual primary production	1985 projected primary demand	1985 projected primary production[1]	2000 projected primary demand	2000 projected primary production[1]
Graphite	ST	*	*	70	0	95	0
Gypsum	ST	14,334	9,436	22,400	10,900	35,000	11,700
Kyanite	ST	*	*	282	223	630	310
Mica, scrap & flake	ST	119	119	231	180	448	227
Mica, sheet	lb.	6,514	0	1,970	0	600	0
Perlite	ST	440	456	770	679	1,350	898
Pumice	ST	3,497	3,132	6,500	5,910	12,000	8,120
Sand and gravel	ST	944,000	944,000	1,740,000	1,450,000	3,200,000	1,900,000
Stone, crushed	ST	648,000	648,000	1,280,000	1,130,000	2,520,000	1,560,000
Stone, dimension	ST	1,740	1,565	2,580	2,030	3,820	1,880
Talc	ST	948	1,028	1,660	1,310	2,900	1,620
Vermiculite	ST	221	285	400	354	725	434
Anthracite	ST	8,248	9,729	5,000	0	2,300	0
Bituminous coal & lignite	ST	517,000	603,000	845,000	612,000	1,000,000	690,000
Natural gas (dry)	CF[2]	21,367	21,015	38,200	31,400	49,000	42,400
Peat	ST	800	517	1,400	1,070	2,200	1,450
Petroleum (inc. nat. gas liq.)	bbl	5,364,000	4,123,000	8,600,000	5,070,000	12,000,000	6,280,000
Shale oil	bbl	0	0	(Shale oil included in petroleum above)			
Uranium, metal	ST	8.320	10.827	51.000	21.100	62.000	28.400
Argon	ST	154	154	390	223	772	329

Helium	CF	587,000	647,000	1,220,000	1,570,000	2,500,000	2,230,000
Hydrogen	CF[2]	2,340	2,340	4,955	3,620	10,500	5,270
Oxygen	ST	13,090	13,090	41,300	24,300	64,500	35,500

Note: Stars indicate information withheld to avoid disclosure of confidential company data. ST = short ton; lb. = pound; kg. = kilogram; t. oz = troy ounce; fl = flask; LT = long ton; CF = cubic feet; and bbl = barrels.

[1] Projections based on the past twenty-year trend, 1951–70. However, the effects of economic, technological, environmental, political, and social factors may alter the supply-demand trends of a specific mineral commodity, and may also affect other commodities where interrelationships are involved. In such cases significant departures from the twenty-year trends could result.

[2] The figures for natural gas and hydrogen are in billions of cubic feet.

Table 3.3. Comparison of Values of United States Primary Demand with United States Primary Production: 1970, 1985, and 2000

Commodity	Price units	1970 actual average unit price	Value of 1970 actual primary demand	Value of 1970 actual primary production	Value of 1985 projected primary demand (at 1970 prices)	Value of 1985 projected primary production (at 1970 prices)	Value of 2000 projected primary demand (at 1970 prices)	Value of 2000 projected primary production (at 1970 prices)
					Millions of 1970 dollars			
Aluminum	lb.	$ 0.29	$ 2,290	$ 338	$ 6,670	$ 284	$ 15,300	$ 292
Antimony	lb.	1.44	42	6	76	1	138	0
Arsenic	ST	168.	*	*	5	—	6	—
Barium	ST	25.13	20	12	25	12	35	11
Beryllium	lb.	60.	*	*	98	7	212	10
Bismuth	lb.	6.00	13	*	18	4	24	4
Boron	ST	510.	45	89	83	130	155	176
Bromine	lb.	.27	93	95	112	123	135	162
Cadmium	lb.	3.57	33	13	59	16	110	16
Calcium	ST	4.03	361	361	592	411	971	476
Cesium	lb.	.60	*	0	*	0	—	0
Chlorine	ST	75.	732	732	1,520	1,070	3,170	1,490
Chromium	ST	154.	71	0	108	0	172	0
Cobalt	lb.	2.20	36	*	44	0	54	0
Columbium	lb.	1.65	8	0	16	0	32	0
Copper	lb.	.58	1,820	2,000	3,360	2,220	6,260	2,760

Fluorine	ST	113.	69	14	137	8	271	5
Gallium	g.	.70	*	*	*	—	1	—
Germanium	lb.	127.	5	4	6	—	10	0
Gold	t. oz	36.41	224	63	335	46	521	36
Hafnium	lb.	85.		0	6	0	9	0
Indium	t. oz	2.50	*	*	2	1	2	1
Iodine	lb.	1.45	7	*	13	1	22	1
Iron	ST	62.	5,210	3,660	7,010	3,160	9,490	3,100
Lead	lb.	.16	265	183	349	134	458	150
Lithium	lb.	.78	*	*	9	7	17	9
Magnesium, metal	lb.	.35	67	78	165	81	406	99
Magnesium, nonmetallic	ST	84.50	89	87	128	119	183	141
Manganese	ST	48.21	64	3	85	0	114	0
Mercury	fl	408.	22	11	27	15	33	18
Molybdenum	lb.	1.72	84	192	166	241	323	316
Nickel	lb.	1.28	399	39	630	77	986	109
Nitrogen, compounds	ST	54.50	565	595	1,110	992	2,160	1,420
Nitrogen, gas & liquid	ST	26.20	139	139	275	223	548	333
Palladium	t. oz	38.	21	—	30	1	40	1
Phosphorus	ST	39.	142	204	257	332	468	456
Platinum	t. oz	133.	54	1	84	0	133	0
Potassium	ST	43.	168	97	295	157	516	200
Rare earths	lb.	.35	8	*	11	15	15	22
Rhenium	lb.	875.	3	5	5	7	11	11
Rhodium	t. oz	215.	8	0	12	—	18	—
Rubidium	lb.	.37	*	0				0
Scandium	g.	4.12	—	0	—	0	—	0
Selenium	lb.	9.	10	9	12	8	15	9

Table 3.3. Comparison of Values of United States Primary Demand with United States Primary Production: 1970, 1985, and 2000 (continued)

Commodity	Price units	1970 actual average unit price	Value of 1970 actual primary demand	Value of 1970 actual primary production	Millions of 1970 dollars			
					Value of 1985 projected primary demand (at 1970 prices)	Value of 1985 projected primary production (at 1970 prices)	Value of 2000 projected primary demand (at 1970 prices)	Value of 2000 projected primary production (at 1970 prices)
Silicon	ST	430.	228	228	323	292	430	380
Silver	t. oz	1.77	129	80	219	69	372	71
Sodium	ST	18.85	383	371	697	509	1,270	679
Strontium	ST	49.	*	0	1	0	2	0
Sulfur	LT	23.15	211	221	382	273	695	336
Tantalum	lb.	8.69	10	0	18	0	35	0
Tellurium	lb.	6.	2	1	2	1	2	1
Thallium	lb.	7.50	*	*	—	—	—	—
Thorium	lb.	7.97	*	*	11	1	24	1
Tin	lb.	1.74	207	*	273	—	351	0
Titanium, metal	lb.	1.32	63	0	172	0	444	0
Titanium, nonmetallic	lb.	.45	419	248	833	405	1,630	518
Tungsten	lb.	2.81	46	23	96	13	208	6
Vanadium	lb.	4.37	62	49	128	87	270	114
Yttrium	lb.	9.	*	*	5	—	8	—
Zinc	lb.	.15	391	160	546	150	900	146

Commodity	Unit							
Zirconium, metal	lb.	6.	*	0	—	0	—	0
Zirconium, nonmetallic	ST	116.	*	*	13	8	20	11
Subtotal			*15,500*	*10,500*	*27,700*	*11,700*	*50,200*	*14,100*
Asbestos	ST	111.	81	14	149	22	270	30
Clays	ST	4.90	260	270	470	343	853	402
Corundum	ST	22.	—	0	—	0	—	0
Diatomite	ST	55.	24	33	52	53	110	69
Feldspar	LT	15.	9	10	19	13	38	15
Garnet	ST	92.	1	2	3	3	5	3
Graphite	ST	45.	*	*	3	0	4	0
Gypsum	ST	3.72	53	35	83	41	130	44
Kyanite	ST	98.	*	*	28	22	62	30
Mica, scrap & flake	ST	21.	2	2	5	4	9	5
Mica, sheet	lb.	.72	5	0	1	0	—	0
Perlite	ST	11.	5	5	8	7	15	10
Pumice	ST	1.14	4	4	7	7	14	9
Sand & gravel	ST	1.18	1,110	1,110	2,050	1,710	3,780	2,240
Stone, crushed	ST	1.58	1,020	1,020	2,020	1,790	3,980	2,460
Stone, dimension	ST	60.80	106	95	157	124	233	115
Talc	ST	7.56	7	8	13	10	22	12
Vermiculite	ST	23.	5	7	9	8	17	10
Subtotal			*2,720*	*2,640*	*5,080*	*4,160*	*9,540*	*5,450*
Anthracite	ST	10.80	89	105	54	0	25	0
Bituminous coal & lignite	ST	6.26	3,240	3,770	5,290	3,830	6,260	4,320
Natural gas (dry)	Mcf	.17	3,650	3,570	6,490	5,340	8,330	7,210
Peat	ST	11.	9	6	15	12	24	16
Petroleum (inc. nat. gas liq.)	bbl	3.18	17,100	13,100	27,300	16,100	38,200	20,000

63

Table 3.3. Comparison of Values of United States Primary Demand with United States Primary Production: 1970, 1985, and 2000 (*continued*)

Commodity	Price units	1970 actual average unit price	Value of 1970 actual primary demand	Value of 1970 actual primary production	Value of 1985 projected primary demand (at 1970 prices)	Value of 1985 projected primary production (at 1970 prices)	Value of 2000 projected primary demand (at 1970 prices)	Value of 2000 projected primary production (at 1970 prices)
							Millions of 1970 dollars	
Shale oil	bbl	—				(Shale oil included in petroleum above)		
Uranium, metal	lb.	7.40	123	160	755	312	918	420
Subtotal			*24,200*	*20,700*	*39,900*	*25,600*	*53,800*	*32,000*
Argon	ST	270.	42	42	105	60	208	89
Helium	Mcf	26.	15	17	38	41	65	60
Hydrogen	Mcf	.25	585	585	1,240	905	2,630	1,320
Oxygen	ST	11.	144	144	454	267	710	391
Subtotal			*786*	*788*	*1,840*	*1,280*	*3,600*	*1,860*
Grand total			*43,200*	*34,600*	*74,500*	*42,700*	*117,100*	*53,400*
Shortfall of primary production compared to primary demand			(8,600)		(31,800)		(63,700)	

Note: Stars indicate values withheld to avoid disclosure of confidential company data, but included in totals; dashes indicate values less than $1 million. lb. = pound; ST = short ton; g. = gram; t. oz = troy ounce; fl = flask; LT = long ton; Mcf = millions of cubic feet; and bbl = barrels.

"consumption" also is a misnomer except from the point of view of the wire mill or brass mill. Most of this new copper is really not "consumed" at all but instead it is converted into copper wire, copper tubing, brass, and alloys from which useful articles having a long life are fabricated. Most of the copper tubing, the copper wiring, the brass hardware, etc., in use now was reported as "consumed" many years ago. But it is estimated that the pool of copper in use in the United States is probably of the order of four hundred pounds per person. Another clear example of the pool of mineral materials in use can be seen in motor vehicles. The more than 110 million motor vehicles registered in the United States constitute a pool of steel in use equivalent to at least two years' full United States demand for steel. When we add up the steel in buildings, railroads, machinery, and so on, it is obvious that the equivalent of many years of annual production is in current everyday use. Some of the material in use could last longer with better maintenance, and some, once discarded, could be more effectively recycled. The same situation applies to almost all mineral materials except the fuels, which are commonly considered to be consumed. However, even fuels are really not "consumed" in many applications but rather the chemical energy originally contained therein is converted to more useful and enduring forms. For example, coal is used to make coke, coke is used to make steel, and the steel then remains in use for long periods of time. Fuel is burned to generate electricity. Substantial

quantities of electricity are used to produce metallic
aluminum. Much metallic aluminum is then used in
enduring applications. Consequently, as we attempt
to forecast future demand resulting from the desires
of other peoples to attain higher standards of living,
we must consider the quantities of minerals that
would be involved, not just the annual per capita
new increments.

No set of future projections can be guaranteed to
be correct. In the past, most projections of mineral
demand have tended, if anything, to be too low, and
programs based thereon have been too modest. But
projections are valuable in indicating the order of
magnitude of the future demand-supply situation,
because any set of reasonable projections of future
United States and world mineral demand make it
evident that long-term demands will require substantial additional mineral supplies. Long lead times are
required to find mineral resources, to convert them to
blocked-out reserves, to mine and concentrate them,
to refine them, and to convert them into energy and
processed materials of mineral origin, while also
conserving, recycling, and rehabilitating. Accordingly,
highest priority must be given to closer industry-
government-academia cooperation and public cooperation at every stage of the complicated process
by which resources are converted into materials and
energy to meet the needs of mankind.

4

Vincent E. McKelvey

Mineral
Potential of the
United States

      Contrary to the popular impression, in part created by the term *nonrenewable resources*, mineral resources are not finite quantities but change over time. Minable reserves are reduced by production, of course, and they may also be reduced by decrease in prices, increase in costs, increased availability of substitute sources, or government regulations that restrict the production or use of various kinds of materials. But they may also be enlarged by new discoveries and by new technologic or economic developments that make it possible to produce from deposits that could not be mined economically before. In attempting to assess the total resources of the United States, therefore, it is necessary to consider not only presently minable deposits already identified but also potential resources, including those of similar quality that are as yet undiscovered and those of lower quality that may some day be economically producible.

For many years geologists have made estimates of probable and possible reserves, but, except for a few commodities, little attempt has been made to take account of undiscovered or presently unworkable resources. For many commodities, the only quantitative resource estimates available are those of proved reserves, which are, of course, essential for many purposes but do not provide a basis for policy decisions concerning future supply.

Recognizing the need for comprehensive estimates of our potential mineral resources, the U.S. Geological Survey is now preparing estimates of United States potential resources of nearly all minerals in present use. Most of the statistical results will be included in the next annual report of the Secretary of the Interior prepared in response to the Minerals Policy Act, but in 1973 they will also be published, along with comprehensive, authoritative reviews of the geology of each commodity, in a four-hundred-page USGS Professional Paper entitled "Potential Mineral Resources— A Geologic Perspective."*

Estimates of potential resources in these studies follow a classification I described early in 1972 in an article in the *American Scientist*,[1] with some improvements added by Donald A. Brobst and Walden P. Pratt (see fig. 4.1). The rationale of this classification is that in order to differentiate known and presently recoverable resources—that is, reserves—from those

*Published in 1973 as U.S. Geological Survey Professional Paper 820, *United States Mineral Resources*.

1. V.E. McKelvey, "Mineral Resource Estimates and Public Policy," *American Scientist* 60 (1972): 32–40.

that are undiscovered, as well as from those that are known but are not now economically recoverable, a classification must convey two prime elements of information: the degree of certainty about the exist-

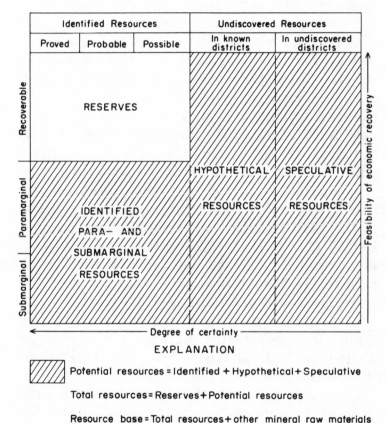

	Identified Resources			Undiscovered Resources	
	Proved	Probable	Possible	In known districts	In undiscovered districts

Figure 4.1. Classification of mineral resources being used by the U.S. Geological Survey in assessing total mineral resources in the United States

ence and magnitude of the materials and the economic
feasibility of recovering them. The diagram in Figure
4.1 shows the classification we are using and the way
in which it takes account of these fundamental
elements of information. Thus, degree of certainty
increases from right to left, and feasibility of recovery
increases from bottom to top. The terms *proved*,
probable, and *possible* refer to *identified* resources—
deposits known to exist. Such deposits are considered
reserves if they can be mined with current technology
and economics; if not, they are considered paramarginal or submarginal. I have defined paramarginal resources as those recoverable at costs as much
as 1.5 times those that can be borne now, and submarginal resources as those conceivably producible
economically in the future at higher prices or with
comparable advances in technology. The key words
in this definition are "conceivably" and "economically." It is hard to conceive that ordinary biotite
granite will some day be mined as a source of aluminum or potash or that ordinary basalt will be mined for
iron. It is conceivable, however, that some day we will
be mining the Revett Formation of the Belt Supergroup for copper, the Laramie anorthosite for
aluminum, the Chattanooga Shale for uranium, and
the Conway granite for thorium.

For the purposes of this volume, we decided that
it would be worthwhile to try to draw a distinction
between resources that we may still reasonably expect
to find in known districts and resources that may be

discovered in geologic terranes or provinces that are broadly favorable but in which there are as yet no discoveries. On the suggestion of Brobst and Pratt, we are referring to these respectively as *hypothetical* and *speculative resources*. To demonstrate that this distinction is useful, let me cite a few examples. In the case of uranium in sandstone-type deposits, we would consider as hypothetical resources the deposits that have so far eluded discovery in the known districts of New Mexico or Wyoming; our geologic knowledge tells us that there is good reason to expect we will find more deposits in those areas similar to those already known, and our estimate of hypothetical resources is, in this case, an attempt to quantify the potential of these undiscovered resources, mainly on the basis of the extent of unexplored but favorable ground. A different kind of geologic perspective, on the other hand, suggests that there is still some likelihood of finding not just new deposits, but new major uranium districts. Our estimate of these *speculative* resources, even though it may be only an order of magnitude, is an attempt to quantify the resource potential of sedimentary basins that are known to exist but have not been sufficiently tested for uranium.

A similar distinction is useful with respect to resources of the base metals. We are justified in *hypothesizing*, for instance, that porphyry copper deposits are concealed under basin fill in and near the known copper districts of the southwest, and the discovery in recent years of porphyry-type copper deposits of Paleozoic and Precambrian age in Eastern North America permits us to *speculate* that whole new regions have some potential for new discoveries.

Another example of the distinction between hypothetical and speculative resources involves the black bedded deposits of barite such as those currently being mined principally in Nevada and Arkansas. The known deposits are scattered over wide areas of both states, and chances are excellent that new deposits— hypothetical resources—will be found. But recent geologic studies of these deposits suggest that they are related more to sedimentary processes than was formerly believed. If this is true, then the potential for the discovery of new deposits and districts in other sedimentary basins—speculative resources—is likewise excellent. This example is an indication of how the reexamination and study of basic geologic principles can help to open new areas and environments favorable for exploration.

One further aspect of the value of such a distinction is that it has forced us to realize that the speculative resources of some commodities are relatively low. From a world-wide geologic perspective we can say that the regions with significant potential for discovery of iron, or phosphorite, or marine evaporites are largely known; thus the significant undiscovered resources fall under the heading of "hypothetical" rather than "speculative." For some commodities this may simply reflect insufficient knowledge of regional geology to identify other promising areas, but for others it is an expression of confidence that the favorable terranes have all been identified.

The entire field in the Figure 4.1 diagram represents total American resources—identified reserves plus potential resources consisting of undiscovered deposits of presently recoverable quality, as well as both identified and undiscovered deposits not recoverable now but conceivably producible economically in the future. Transfers or conversions of potential resources to reserves are made by discovery, technologic advance and changes in economic conditions. And the same processes may lead to similar transfers from presently unknown sources in the resource base.

Whereas estimates of reserves change significantly over short periods of time, well-made estimates of potential resources are likely to change more slowly, for they are less influenced by changing economic, technologic, and political conditions. It should be understood, nevertheless, that estimates of potential resources represent neither a forecast of ultimate production nor a total inventory of resources ultimately producible. Ultimate production will depend not merely on the geologic availability of raw materials, but on many other factors, such as progress in the technology of exploration and recovery, development of substitute materials or more competitive sources, and changes in government regulations or environmental trade-offs that affect the availability of land for exploration and the producibility of individual

deposits. And our ideas of what is conceivably dis-
coverable and producible will surely change also
with development of new knowledge about the origin
and occurrence of minerals and advances in the
technology of mineral recovery and use. Until a few
years ago we would have made no estimate of copper
in the Revett Formation, because we did not know
that it was copper-bearing or even that that kind of
copper deposit existed in the United States. And we
would not have attempted to estimate potential
resources of germanium or scandium because they
had no known uses.

Rather than an inventory of a fixed supply or a
forecast of ultimate production, estimates of potential
resources are state-of-the-art-and-knowledge apprais-
als of the geologic availability of raw materials
presently judged to be of future commercial interest.
They will change with time, but meanwhile they are
valuable in establishing targets for exploration and
for research in mining and recovery technology, and
they are valuable also in identifying problems and
alternatives for policy decisions on the part of government.

Some of the results of this mineral resources appraisal program are reported in Figures 4.2 and 4.3 in the
form of graphs that compare our estimates of identified, hypothetical, and speculative resources of several
minerals in the United States with the Bureau of
Mines' forecasts of the total cumulative demand for

these minerals for the period 1968–2000.[2] The upper bar of each pair shows the bureau's estimated range of cumulative demand for the mineral; the lower bar shows our estimates—or in some cases, the bureau's estimates—of present reserves, and then, cumulatively, our estimates of identified para- and submarginal resources, hypothetical resources, and (where quantitative estimates have been made) speculative resources.

Shown on Figure 4.2 are the estimates for iron and some of the ferro-alloys. Obviously the United States is in fair shape for iron—reserves are somewhat more than the maximum cumulative demand to the year 2000. The tremendous identified paramarginal and submarginal resources of iron ore are mainly in low-grade Lake Superior ores that require beneficiation and agglomeration.

The United States is also well fixed for molybdenum. The identified para- and submarginal resources of molybdenum are mainly in porphyry molybdenum and copper-molybdenum deposits at grades lower than present cutoffs.

Reserves of vanadium and tungsten are much less than projected demand through the year 2000. Identified resources of vanadium in magnetite deposits and in carbonaceous shales are more than adequate to

2. U.S. Bureau of Mines, *Mineral Facts and Problems*, Bulletin 650 (Washington, D.C.: U.S. Government Printing Office, 1970).

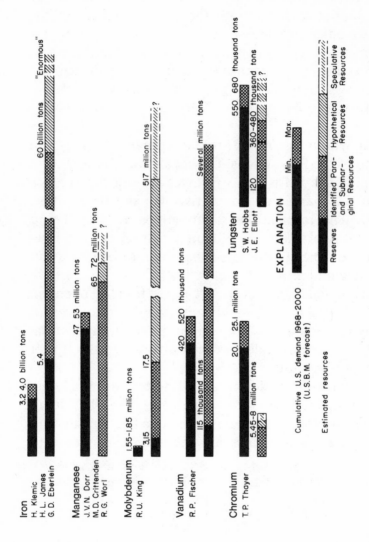

Figure 4.2. United States resources of iron and ferro-alloys compared to projected demand through year 2000. Names beside each resource bar indicate the Geological Survey geologists who prepared the individual estimates.

make up the deficiency, but they can do so only if domestic metallurgical practices are modified or new ones are developed. For tungsten, identified para- and submarginal resources make up a large part of the deficit. There is some additional prospect, not quantified here, of by-product recovery from the known association of tungsten with the porphyry molybdenum deposits.

For the remaining ferro-alloy elements shown here, manganese and chromium, the United States has no domestic reserves. Identified submarginal resources of manganese are large but are both very low-grade and refractory to economic concentration. The principal hopes of finding domestic reserves or resources lie in (1) finding the source of the manganese in the Pierre Shale, conceivably buried under Pleistocene sedimentary rocks in central or western Minnesota or adjacent areas; (2) finding another Molango-type deposit by careful analysis of the distribution of manganese in certain miogeosynclinal carbonate rocks; or (3) finding the source of the high manganese concentrations in the Salton Sea brines. Domestic identified resources of chromium are low-grade and represent only about a four- to five-year supply. Moreover, the outlook for substantially increasing the domestic resource base, even at several times present world prices, is not favorable. The thicker, minable parts of the Stillwater Complex in Montana are complexly faulted and sheared off at depth. Concealed podiform deposits in the Appalachians or along the

Pacific Coast offer little promise. The principal potential North American resources other than the Stillwater are a similar complex in Manitoba and very large low-grade deposits in Greenland.

Turning to nuclear energy (Figure 4.3), *identified* uranium resources in conventional deposits are less than half the minimum projected requirements to the year 2000, and even the hypothetical and speculative resources of conventional type barely make up the deficit. Domestic *reserves* are sufficient to last about another decade, but needs beyond then are so great that tremendous efforts in exploration and research in ore-finishing techniques will be required to discover and develop new recoverable resources. The real long-range promise here seems to be the low-grade resources in marine phosphorite; the identified para- and submarginal resources alone contain over twenty times the uranium in present reserves, but to obtain significant supplies of uranium from these sources would require mining and treating vast quantities of rock, disrupting large areas of ground at high unit costs.

The thorium picture is more complicated. First, the forecast range between minimum and maximum demand is noticeably greater than for other commodities; the maximum demand assumes successful development of a commercial thorium reactor by 1980. Reserves—recoverable by-products from Atlantic Coast beach placers—are not adequate for the maximum projected demand. Identified resources are huge;

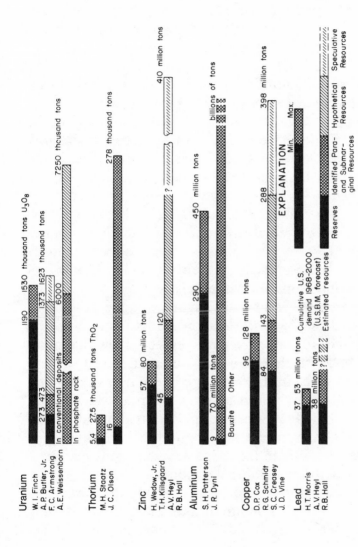

Figure 4.3. United States resources of uranium, thorium, and major nonferrous metals compared to projected demand through year 2000. Names beside each resource bar indicate the Geological Survey geologists who prepared the individual estimates.

about 40 percent of these are in relatively high-grade vein deposits, but the balance are in low-grade conglomerates and fluviatile placer deposits.

Lead reserves are about equal to projected demand to the year 2000, and the outlook for the continuing discovery of additional reserves and resources of lead at a rate that exceeds consumption is believed to be excellent; para- and submarginal resources in the United States have not been quantified but probably are of at least the same order of magnitude as the reserves. Many of the undiscovered deposits are assumed to be comparable in tonnage and grade to known ore bodies. Zinc reserves are also substantial, and undiscovered zinc resources in the United States are thought to be mainly in strata-bound deposits similar to those in carbonate rocks in the Mississippi Valley and Appalachian regions.

Reserves of copper are slightly less than the minimum of the forecast demand range, and the total identified resources slightly exceed the maximum. The bulk of the domestic potential resources are believed to be in the western conterminous states and Alaska—including an apparently large resource potential in the Belt Supergroup. As I mentioned before, however, the recent recognition of porphyry-type copper deposits in the Appalachians opens up whole new potentially favorable regions.

At present the United States imports nearly 90 percent of its aluminum, as alumina and bauxite; the remainder is produced from domestic bauxite deposits, and both the reserves and an additional 60 million tons of identified para- and submarginal resources represent domestic bauxite. The enormous identified resources beyond this are contained in a variety of other aluminous materials that pose technological problems; the most promising of these are high-alumina clays and dawsonite in the rich oil shale deposits of Colorado. The dawsonite alone, in one area of two hundred fifty square miles, contains an estimated identified resource of some three and a half billion tons of aluminum.

These estimates illustrate several conclusions that can be drawn from an analysis of United States potential resources. For some commodities such as manganese, tin, and chromite, the United States still must look to foreign sources for future supplies. For materials such as vanadium and tungsten future domestic production depends on advances of extractive technology or substantially higher prices that will permit the use of lower grade resources. However, resources of materials such as iron, molybdenum, copper, lead, zinc, and aluminum are nearly equivalent to potential demand over the next few decades, and the prospects for new discoveries are reasonably good, provided that exploration is pursued vigorously.

The philosophy that emerges from these considerations is that ample raw materials are indeed there in the ground—they are geologically available. Innovative application of old geologic theory and the creation of new concepts of ore-formation can be expected to lead to the discovery of minable deposits, and past experience tells us that research and exploration can be counted on to discover some kinds of deposits and some ore environments that we do not know about now. The bertrandite (beryllium) and dawsonite deposits are examples of recent discoveries of new kinds of deposits, and the Revett Formation is an example of the new kind of ore environment. Many examples could also be cited of the technological advances that we can expect to move para- and submarginal deposits into the recoverable category. In spite of immediate environmental and technological problems that face the United States, I am confident that we can shift the para- and submarginal identified resources and undiscovered reresources into the reserve corner of Figure 4.1 and add potential resources from the resource base as well—provided we are both aggressive and imaginative in our pursuit of research and exploration.

5

David Swan

Science
and Technology:
Aid and Hindrance

One of the first things I did when asked to write a chapter on the subject of science and technology was to reread the President's Minerals Policy Commission report (the Paley report), dated June 1952. In the volume dealing with *The Promise of Technology*, six major areas of technological need were identified. Twenty years later they are still a good listing of the technical needs for the minerals segment of our society as it is, and as it is likely to be during the remainder of the century.

The technical problems requiring solution, as identified by the Paley Commission in 1952, were (1) to foster new techniques for discovery; (2) to bring into the stream of use materials which so far evade our efforts; (3) to apply the principle of recycling more and more broadly; (4) to learn how to deal with low concentrations of useful materials; (5) to lessen or eliminate the need for a scarce material by substituting one that exists in greater abundance; and (6) to

develop and use more economically the resources
that are recoverable in nature.[1]

While the Paley Commission's scope was broader
than that of this volume, since it included materials for
clothing and thus was concerned with nonminerals, I
believe it adequately identifies the major areas where
technology may play a significant role in altering the
mineral position of the United States during the next
quarter century.

A Frame of Reference for Minerals Use

In considering the broad impact of science and technology on future United States material supply and
demand, it is helpful to provide an analytic framework
so that the various forces acting on future supply and
demand can be better identified. In such a conceptual
framework we can visualize minerals as providing
the basic *materials* from which can be constructed
fabrications, such as bars, sheets, billets, etc., which
in turn are converted into *components* by machining,
welding, casting, and so on. The components are in
turn assembled into *subsystems*, such as automobiles
and airplanes, which form key parts of *systems*, such
as the air transportation system, all of which perform
basic *functions* necessary to society, such as shelter or
transportation.

1. President's Materials Policy Commission, *Resources for
Freedom*, vol. 4, *The Promise of Technology* (Washington, D.C.:
U.S. Government Printing Office, 1952).

At each level of conversion, one acquires a larger number of choices in selecting the particular set of materials capable of fulfilling the requirements of the particular segment being considered. All of the various translations require large energy inputs, so that energy supply and availability predetermine the number of options at each level.

Like any model, this is vastly oversimplified in that there are feedback effects which affect the behavior of each segment, particularly in a highly industrialized, interdependent society such as ours. However, the model does enable us to play "what-if" games as an aid to identifying promising areas of science and technology which might affect future materials supply and demand.

To illustrate some of the profound changes which can occur at the functional level, we can consider the ascendancy of the jet transport for long-distance transport of people. It has largely eliminated the transoceanic steamship as a mover of people, relegating it to a role of pleasure cruising, and it has effectively eliminated the transcontinental passenger train system. One could, in principle, examine the effect of these changes on the demand for minerals, although a thorough comparison of demands of the various transport systems would require examination of such second-order effects as the need for steel for the tankers (or pipe lines) which transport the additional fuel burned by the jet plane.

Economists have for many years been wrestling with

the kinds of problems alluded to above, although their interest has been focused on including the effects of technical change on their forecasts rather than in trying to identify the technologies most likely to bring about change. The development of large-scale computers has added a new tool to this effort. However, successful, broad application of the techniques for forecasting discontinuous technical (or economic) change has so far eluded its proponents, due largely to the unavailability of the proper data to feed the computer.

Having identified a few of the difficulties in attempting to get at the complex problems of demand forecasting, particularly for a twenty-five-year look ahead, let me now attempt to identify the key factors affecting the United States demand for minerals and see if anything useful can be gleaned from the very imprecise economic theory and data which are now available to us.

All long-range demand forecasts that I am familiar with start with the plausible assumption that demand is somehow related to population growth. Since the time of Malthus, economists (and demographers) have been attempting to calculate that moment in time when people will occupy the entire earth's surface. As a matter of interest, world population is now growing at an annual rate of 2.1 percent, indicating that if this is sustained, the inhabitable land area of the earth will be occupied by people lying down, like sardines, in the year 2205, only 233 years from now. If we assume that each person needs a support area (for food and shelter) of about one-half acre, all arable land will be consumed by 2007 (34 years hence).[2] It therefore seems reasonable that the exponential growth curves for population which have been observed for the last two hundred years or so represent the portions of typical S-shaped growth curves which have been observed for many forms of organic growth. One possible scenario, then, is to postulate a leveling off of population growth to perhaps linear or even zero growth. Some confirmation of this appears in western European and more recent United States birth rates.[3]

Many projections of material demand have been made, using the same general assumptions. In the recent report of the Secretary of the Interior to the Congress, pursuant to the National Minerals Policy Act, United States demand for aluminum was projected as 26,400,000 tons by 2000, while world demand (assumed to grow at 6.6 percent annually) would be in excess of 60,000,000 tons.[4] If this growth rate were

2. G. W. Perbix, "Perspectives of Recycling" (Remarks delivered at the Annual Meeting of the Bureau de la Récupération, Cannes, France, June 2–5, 1972); reproduced in pamphlet form by General Services Administration, Washington, D.C.

3. I am indebted to Dr. G. W. Perbix of the National Commission on Materials Policy for calculations of population growth.

4. U.S. Department of the Interior, *First Annual Report of the Secretary of the Interior under the Mining and Minerals Policy Act of 1970* (Washington, D.C.: U.S. Government Printing Office, 1972).

sustained to 2140, the entire United States would be covered with a four-inch layer of aluminum. I think we can agree that indefinite projection of exponential demand leads to an unlikely future; therefore, the question before us is to identify timing and effects of a transition from exponential to linear demand. Without attempting to predict exactly what will happen, let us examine the effects of changing growth patterns.

Recycling

One of the first effects of a change from exponential to linear growth is to alter the relative importance of primary versus secondary sources of the material in question. If we assume that all material in use is economically recoverable, then in principle the life cycle of the artifacts containing the material will determine when it is available for recycle. The amount of scrap available would depend on the amount consumed during past years. Thus, the higher the past growth rate of demand, the smaller will be the scrap reservoir relative to current demand. If we now slow down the growth rate in demand, then a larger fraction of current demand can be satisfied from the scrap inventory (less losses due to dilution, preservation as antiquities, and so on). Economic costs of recycling will, of course, determine the amount of the theoretical scrap inventory that is actually recycled versus that provided from primary sources. Since it appears that several of our basic mineral raw materials have actually reached the linear growth stage, it would appear that the technology of recycling will receive more attention in the next twenty-five-year period than in previous times. Environmental concerns seem likely to have little effect on total amount recycled, since the present recycling system for most mineral products is quite efficient. The high degree of concern with recycling municipal wastes, has, however, encouraged considerable research and development work on this problem.

Energy Efficiency of Recovery Processes

The low cost of electric and fuel energy in the United States has encouraged the use of relatively energy-intensive processes for extraction of metals from ore. If, as seems likely, the cost of energy increases, then clearly present processes must be reexamined. More energy-efficient processes must be devised, both because of the direct costs of energy and also because of the high costs of disposing of combustion by-products, that is, waste heat and air pollutants such as sulfur oxides, carbon monoxide, nitrogen oxides, and photo-chemical oxidants. The effect of these trends on technology will be to cause reexamination of extraction processes using chemical rather than thermal energy. We can therefore expect an era of increased consideration of hydrometallurgical processes, the development of reducing agents other than carbon, and some relocation of processes toward energy sources and environmentally acceptable locations.

Air and Water Pollution

The public recognition that the assimilative capacity of air and water resources may be locally overloaded has led to stringent regulations regarding the industrial use of these resources. Obviously, the costs associated with meeting the new regulations act as a strong incentive for process improvement. The problems associated with changing air and water pollution regulations may be divided into two general classes, the first having to do with adding the necessary process changes to enable existing processes to comply with regulations, and the second having to do with choosing the best process technology for future demand. Again, the growth in demand will have a profound effect on the technology chosen. If growth is low, then the primary goal tends to become fixing up existing plant and equipment to comply with new regulations, process replacement being usually an economically unattractive solution, while in high growth industries, new processes can be economically considered for expanded capacity. Thus those extractive industries with low growth rates may be more adversely affected than those with higher growth rates.

Intermaterial Substitution

The existing trend toward development of material systems, or composites, which enable development of a variety of properties more economically than use of a single material, will doubtless continue. While this undoubtedly results in more efficient use of materials, it also complicates the problem of recycling, since the use of multiple component subsystems usually adds to the problem of separating the components when their useful life is ended.

The increasing freedom for large-scale substitutions of basic commodities caused by the larger array of choices available to the eventual consumer is also likely to continue. Thus the major commodity minerals and metals—iron and steel, aluminum, copper, lead, and zinc—will all continue to be faced with threats and opportunities brought about by changes in functional requirements for the end products which utilize them. Requirements for the alloying metals are largely influenced by these same considerations. Finally, the continued development of improved properties in plastics and other organic compounds is likely to continue to affect the consumption of both heavy and light metals during the remainder of the century.

New Materials

The application of knowledge from the science of physics has had a profound impact on the need for new materials. Particular areas are in the nuclear field, which has caused the development of substantial new materials industries in development of reactors and fuel reprocessing. For example, large-scale production of zirconium, uranium, and transuranium elements, as well as high purity graphite, are all the result of this need. Similarly, the revolutionary changes brought about by substituting electronic solid state devices for

thermionic and mechanical devices has spurred the
need for quantities of formerly rare minerals such as
gallium (for light-emitting diodes), high purity silicon,
high purity niobium, and many others. While the
tonnages are not high, the demand will likely increase
rapidly, and new extractive and conversion technologies will continue to be developed.

New Techniques for Discovery and New Sources of Minerals

Others in this volume are discussing application of
new exploration techniques as well as new sources of
minerals which may be available during the last part
of the century; hence I will only mention a few specifics
where the development of new technology may in
time alter our exploration and exploitation base.

The development of space technology may have a
significant impact on terrestrial mineral exploration.
A large-scale cooperative experiment now under way,
the Earth Resources Satellite, will provide additional
insight into possible techniques which may be used.
The problem of precisely and economically deter-
mining underground ore locations, extent, and grade
continues to be elusive, although the computer enables
large amounts of data to be efficiently and economically processed.

The increasing recognition that minerals present in
the deep oceans may be economically recoverable is
another technical challenge of the 70s and 80s. So-
called manganese nodules, many of which contain

significant amounts of copper, nickel, molybdenum, and cobalt, are a potentially abundant resource. The major technical obstacle to their exploration lies in the development of reliable mining or harvesting techniques which will operate in the 15,000- to 20,000-foot deep ocean environment.

Summary

In conclusion, from this author's vantage point, it would appear that the technology of mineral extraction will continue to follow its evolutionary path of the last quarter century. It would appear that for certain basic materials the period of rapid United States growth in demand may be ending, and that the technology necessary to support more recycling must be developed. Finally, the energy consumption of extractive processes must be economized, and the air and water pollution problems, to a large extent associated with them, must be lessened. This has necessarily been a brief summary, more concerned with underlying trends than with specific forecasts, and I hope it will be read in that context.

6 *Peter T. Flawn*

Impact of Environmental Concerns on the Mineral Industry, 1975–2000

The effect of the mineral industry on the environment has been the subject of much investigation, analysis, and action, and not a little rhetoric. What I propose to do is turn the issue around and assess the effects of environmental concerns—including legislation and public policy decisions—on the mineral industry.

These effects are now being experienced by the industry in its many fields of endeavor, and between now and the year 2000 they will constitute a major, if not *the* major, factor in industry operations. Now, and for the next few years, public concern about environmental quality and the public policies resulting therefrom will have a generally negative or depressing effect on the industry (and in some specific cases, a disruptive effect) except in the area of research and in enlarged markets for certain commodities. In my opinion, the ultimate consequence will be most serious in mineral exploration. The problems faced by

(1) the extractive element of the industry, and (2) the processing, beneficiating, smelting, and refining elements of the industry will receive the most attention because of inhibiting regulations and markedly increased planning, capital, and operational costs. But the consequence of direct impacts on these segments of the industry—lowered net earnings and reduced debt service capacity—will be a cut in exploration budgets. Combined with reduced financial capacity to conduct exploration and reduced incentive to look for domestic properties, there is the prospect of reduction of available exploration acreage in the United States through institutional restraints and higher acquisition costs. The reduced exploration thrust will, in the long run, have more serious consequences for the United States than the closing of a smelter or higher reclamation costs for surface mines.

I propose now to look at those several areas where environmental legislation, policies and attitudes have affected and will continue to affect the mineral industry.

Denial of Access to Public Lands and Waters: Increased Costs of Exploration Rights

It is well known to exploration geologists and engineers that significant concentrations of certain minerals and elements useful to society are the products of rare crustal environments of the geologic past, and today the search for those parts of the earth's crust where such concentrations took place requires every scientific and engineering skill that society can command. Compared to the area of the continental land masses of the world—including their shelves—the area underlain by valuable mineral deposits is minuscule—certainly less than one-half of one percent of the total area.

In recent years there has been a move to deny to mineral exploration large areas of public lands through (1) creation of wilderness areas; (2) the use of provisions of the National Environmental Policy Act (NEPA) to delay public land lease sales and drilling permit applications; (3) cancelling of or refusal to grant mineral patents; and (4) (difficult to document) harassment and intimidation by government officials at many levels. The generation of environmental data, the preparation of impact statements, revisions of applications, statements in defense of proposals or actions, and litigation all cost money. At the same time, costs of acquiring exploration rights (bonuses, rentals, royalties, and options) on tracts of public and private lands have risen. I see no prospect for a turn-around in these trends in the near future. Exploration and development of mineral deposits on public lands will continue to meet political resistance and to face higher costs.

Higher Costs for Planning and Operations: Increased Capital Costs

In both the extractive phase and in processing, beneficiating, and refining, environmental concerns are increasing costs. Environmental impacts must be anticipated and dealt with. Reclamation plans must be made, subsidence must be controlled, mine wastes must be disposed of, and equipment to monitor and control effluents and emissions must be installed. If environmental standards cannot be met, projects must be abandoned and facilities must be closed. Clearly these costs will fall more heavily on some mineral industries than others, but all will have to operate within a new economic and engineering framework.

It is difficult to make dollar estimates of these additional environmental costs over the next twenty-five years or so. Early in 1972 the Nixon administration released the results of a joint study by the Council on Environmental Quality, the Department of Commerce and the Environmental Protection Agency that attempted to assess the impact of pollution controls on the United States economy.[1] Eleven industries with major pollution problems were analyzed in detail—five of these were mineral-based industries and included cement, iron foundries, metals smelting and refining (aluminum, copper, lead and zinc), petroleum refining, and steel-making. In these five industries it was forecast that between 1972 and 1976 capital costs for pollution control would be between $4.5 and $6.0

1. *The Economic Impact of Pollution Control—A Summary of Recent Studies Prepared for The Council on Environmental Quality, Department of Commerce, and Environmental Protection Agency* (Washington, D.C.: U.S. Government Printing Office, 1972).

billion and annual costs would rise from about $100 million in 1972 to $1.5 billion in 1976, an overall fifteen-fold increase. These costs are minimum costs based on secondary treatment for water, federal air quality standards, not including solid waste disposal costs, and ignoring inflation. Some states have already imposed tougher air quality standards, and higher water quality standards are threatened. The cost data were generated by the Environmental Protection Agency.

Of the 12,000 plants in the eleven industries studied, the report estimated 800 closings in the normal course of business over the 1972–76 period, with an additional 200–300 closings because of air and water pollution abatement requirements. (An estimate of how many new plants will *not* be built because of these requirements was not included.) Job loss was estimated at 50,000 to 125,000 jobs; about half of the closings are expected to produce a "community impact." Indirect economic impacts (on suppliers and service companies, for example) were not evaluated.

The report predicted that the cement industry would be forced to make changes in its financial policies, including "a reduction in the dividend payout ratio . . . . and an increase in the debt/equity ratio . . . ." The forecast was that about twenty-five plants will close in the four-year period. They will be replaced by large, modern facilities. A 4 to 5 percent real price increase is expected.

Iron foundries will be hard hit. It was estimated that by 1980 pollution abatement will result in the closing of about 400 foundries with loss of 16,000 jobs. There will be price increases.

In the nonferrous metals smelting and refining industry, aluminum is expected to escape without any plant closings, but there will be price increases. The copper industry will lose at least two smelters, with a significant community impact. There will be an increase in domestic prices, but competition will hold down foreign prices. Some high-cost lead and zinc producers will be forced to close, but price increases due to pollution abatement costs will be small.

As many as twelve small petroleum refineries may be forced to close, and these closings in or near small communities will produce a significant local impact. There will be price increases. In steel-making it was forecast that there will be price increases, but no plants forced to close because of pollution abatement requirements.

In summary, the report predicted a general negative effect on the economy that will show up in reduced G.N.P. growth, higher unemployment, and a less favorable balance of payments. Some firms will earn lower profits, others will curtail production, and some firms and plants will be forced to close. However, considering offsetting factors such as increased federal expenditures and economy-stimulating interest-rate policies, these losses were considered "manageable."

It was concluded that none of the eleven industries studied would be hurt severely.

I find this reassurance singularly unconvincing. Unlike usual private-sector capital expenditures for new facilities to expand production or to reduce costs, these expenditures for pollution control have a negative rather than a positive effect on rate-of-return on investment. Capital expended for pollution abatement will not be available for other ventures necessary for the economic health of the industry and the nation. The "good" is a social "good"; the rate-of-return is measured in terms of improved environmental quality. Some mineral industries will be hurt "severely" and losses to the nation will not be "manageable" unless we establish a policy now that will build in the "offsetting factors" and provide a balance; a policy that will distribute the costs of environmental quality beyond the mineral industry, or the chemical industry, or the paper industry.

Smelters have received much publicity because of their problem in meeting state and federal air control standards. Thirteen of the nation's fifteen major copper smelters are in the western states; seven are in Arizona. As a result of new federal air quality standards and tougher standards adopted or contemplated by some states, some smelters have had to reduce operations and some may be temporarily or permanently closed. There is now no significant excess smelting capacity in the American copper industry.

At the very least, the major copper producers are faced with large capital investments to bring smelters into compliance. Phelps Dodge anticipated it would require $240 million to bring its three Arizona smelters into compliance with the 90 percent sulfur recovery standard and that the old Douglas smelter could not be saved under such a rigorous quality standard. Similar statements were made by Asarco with regard to their Tacoma smelter—that 90 percent recovery of sulfur was beyond their technical capability. Kennecott predicted that under such a standard their entire Utah Copper Division would close. Newmont is spending $30 million at San Manuel, Arizona; Anaconda is spending $26 million in Montana; Asarco is spending $20 million at El Paso, Texas. Phelps Dodge has announced a new $100 million smelter in New Mexico. These are only examples, not a complete listing of capital requirements to bring smelters up to environmental quality standards. Investments of this magnitude in smelter improvements and modernization will affect other areas of company operations, particularly exploration and new property acquisition.

Changes in Market Structure for Minerals and Metals as a Direct Result of Environmental Concerns

Some metals and minerals that have been used by man for millennia and have been recognized as toxic or harmful to health for almost that long, have been found to be more widespread and thus more dangerous to health than previously recognized. Analysis of food, air, and soils has disclosed mercury and lead in dangerous or potentially dangerous quantities where it had not previously been suspected. As a direct result, the market for both mercury and lead is threatened. The chlorine-caustic soda industry, previously a major market for mercury (23 percent of the national market), is moving away from the mercury cell to a technology that does not use mercury. The issue of leaded gasoline has been much publicized. About 20 percent of the lead sold in the United States in recent years was for gasoline antiknock compounds. Lead is also in trouble as a paint additive. The mundane use of salt for removal of ice on pavements will be reduced or eliminated. The sulfur market has recently suffered from an excess supply. In the future, air quality standards that require recovery of nearly all sulfur in stack gases or precombustion desulfurization of fuels will make a larger percentage of recovered sulfur available, and the pressure on price will be strong. Asbestos has been attacked as a health hazard, as the tiny fibers are released to the environment from brake linings and poorly packaged home insulation. So far, the effect on the market has been negligible, but it will probably increase.

However, all of the environmental pressure is not negative. Certainly it is reasonable to predict increased demand for certain catalysts, such as platinum (and possibly vanadium, copper, nickel, cobalt, and cadmium), that will have increased use in pollution control devices. The use of expensive catalysts, however, will probably not continue as technology works to find substitutes under a strong cost incentive. Some industrial minerals used as filters and in ion-exchange processes will also enjoy increased demand. New metallurgical technology designed to minimize smelting and maximize hydrometallurgy and electrometallurgy will also result in changes in markets for fluxes, scavengers, and other pyrometallurgical materials. More efficient scrap recovery and use, possibly encouraged by tax incentives, will make a market impact.

In the area of fuels, the eventual balancing of environmental costs and benefits will affect the oil shale, coal, petroleum, natural gas, and uranium industries. The standards that are ultimately adopted for air and water quality, including thermal standards, the legal and economic constraints placed on strip mining, and the legal and economic constraints placed on management of radioactive wastes will determine how the United States uses its energy resources. Hopefully, the direction will come from a carefully constructed national energy policy.

Probably the next quarter century will see a thrust away from combustion. In the case of vehicles the successful development of a practical battery technology would provide a major stimulation of the market for the winning metals. Those now under consideration include sodium (sulfur), lithium, lead, zinc, nickel, cadmium, manganese, mercury, and silver. Mining, beneficiating, and refining these metals present different kinds of environmental problems; therefore the environmental cost factor will not be equal.

Stimulation of Research

What has not yet become generally apparent and what cannot yet be measured is the magnitude and direction of research directly stimulated by environmental concerns. Some efforts, such as development of methods to clean up oil spills and remove sulfur from stack gases, the development of antipollution devices for automobiles (including an electric car), the revegetation of strip mine spoil, and recycling of solid wastes, have received national attention in the press, but other environmentally generated research on mineral industry problems has not been well publicized. Areas that will receive research impetus from environmental legislation, policy, and attitudes over the next twenty-five years are: (a) metallurgy, (b) *in situ* mining, (c) underground excavation, and (d) waste recycling and scrap recovery:

a. Research in metallurgy will be directed toward elimination of smelting through solution chemistry or hydrometallurgy. In the case of copper, for example, flotation concentrates will be treated by chemical processes to produce cathode-grade copper.

b. In the area of mining, basic procedures date from the Neolithic age: dig a hole, break off a piece of valuable rock, put it in a bucket, and hoist it to the surface. Processes to selectively render mobile the desired minerals or elements of an ore body and recover them through boreholes are called *in situ* mining, solution mining, and chemical mining. *In situ* mining is a broader term because it includes mobilizing elements by melting (for example, sulfur) as well as by solution. Solution mining includes industrial minerals (salt) as well as metals. Chemical mining applied to extraction of metals is in the broad sense part of hydro-metallurgy. Solution or *in situ* mining is not without environmental problems involving groundwater con-tamination, the disposal of large quantities of soluble salts, and subsidence of the surface. However, it avoids large volumes of mine wastes and mill tailings and, in the case of ore bodies within one thousand feet of the surface, it avoids the great scars of open pit mines.

c. The concerns about environmental damage from strip or open pit mining have prompted research designed to increase the rate and reduce the cost of underground mining. This research has perhaps received even more impetus from the needs of urban areas to efficiently utilize the subsurface for trans-portation and utility systems. A report in 1968 by the National Academy of Science, National Academy of Engineering, and National Research Council on *Rapid Excavation: Significance, Needs, Opportunities* pointed out that surface excavation technology is far in advance of underground excavation technology and that tunneling of rock needs the greatest attention. As a goal, the report called for a 30 to 50 percent reduction in real costs of underground excavation and a 200 to 300 percent increase in the sustained rate of advance.

d. The next quarter century will see continued research to develop new technology for scrap and waste collection, handling, recovery, product-development, and use. It will be encouraged through various kinds of incentives in the form of special tax treatment or subsidies. As these new scrap and waste systems come into use they will have a significant effect on metal and mineral markets.

In conclusion, the mineral industry is in for a period of very rapid change over the next twenty-five years. All segments of the industry—exploration, mining, beneficiating, refining, transportation, marketing, and research—will be affected. The industry must have the financial capability to respond and adapt to change.

There will be increased costs. Those affected will attempt to recover increased costs through price increases. If the price increases are permitted, consumers of mineral products will turn to cheaper foreign sources, unless prevented by import controls. If we increase the flow of foreign mineral materials and products the domestic industry will suffer, and the balance of payments will be adversely affected.

If price increases are not permitted, there may be for a time an intensive research and development effort to reduce costs. If this is not successful, and there is no federal subsidy or loan program to provide relief, the industry will decline. If environmental quality standards are unattainable, the industry will face disruption. All of these alternatives suggest a trend toward a decrease in per capita consumption of mineral products. The magnitude, direction, and consequences of these trends can be quantitatively evaluated over the next few years.

It seems clear that without an enlightened national mineral policy that defines the interests and objectives of the nation and sets out specific actions required to accomplish those objectives, United States industry will face potentially destructive supply problems in obtaining mineral raw materials, including mineral fuels, and potentially destructive economic problems in maintaining operations. Indeed, the economic impact will extend far beyond the mineral industry. The solution to the anticipated problems lies in the public policy area.

7

Stanley Dempsey

Land Management and Mining Law: The Framework of Mineral Development

Concern for the natural environment, first expressed as a collective national concern at Earth Day observances in 1970 and more recently expressed as a global concern by the United Nations Conference on the Human Environment held at Stockholm in June 1972, has focused great attention on questions relating to mineral supply and environmental impacts associated with mineral-producing activities. Problems of environmental degradation are now better understood. Shortfalls in mineral supply, present or potential, are widely discussed, and "energy crisis" is a part of our language. As efforts are made to find solutions to environmental problems, there appears to be an emerging consensus that modern institutions are ill equipped to deal with the problems generated by massive technology and that ultimate solutions will require sweeping institutional rearrangement. Changes in legal institutions may have profound effects upon

mineral-producing capacity, and the ability of the
United States to achieve its mineral supply goals
during the period 1975–2000 will be directly influenced
by such institutional rearrangement.

This chapter is addressed specifically to land management and mining law as elements of the larger
institution of law in general and to the role of mining
law in the development of mineral supply strategies.
Mining law will be defined broadly, and the chapter
will briefly trace the sources of such law. Emphasis
will be given to the policy basis for mining law, and
how such law has responded to policy shifts in the past.
Specifically, it is my intention to show that the legal
framework of mineral-producing activities can be
adjusted to help achieve the nation's mineral supply
goals, but that to accomplish such adjustment we
must: (1) adopt clear national minerals policy goals,
(2) avoid temptations to use mineral tenure laws to
achieve "other" policy goals, and (3) eliminate incon-
sistent governmental policies and the accompanying
confusion of overlapping statutes, rules, and enforcement procedures. Much of this chapter deals with the
land law framework of domestic minerals production,
but the principles discussed also bear upon land problems involved in developing off-shore mineral supplies.

Mineral Lands Use Law

"Mining law" is often thought of narrowly as
dealing only with the rights of miners or with public
land mineral disposal statutes. The subject is much

broader. It would be more descriptive to use the term "mineral lands use law."

Viewed jurisprudentially, mining law can be considered to encompass rules concerning the ownership of mines and the relationship between miners and other users of lands, together with the enforcement of those rules by action of the state. The rights and duties which these rules prescribe make up the framework for mineral development as we know it today.

Broad categories of mining law rules include:

1. rules permitting private parties to acquire for mining purposes lands owned by the state;

2. rules providing for the resolution of disputes between rival claimants where lands of a state are available for acquisition on a nonexclusive basis;

3. rules concerning the conduct of exploration and mining on lands where the mineral and surface estates are owned by different parties;

4. rules dealing with the impact exploration and mining may have on adjoining lands which are owned by others;

5. rules dealing with the impacts of land use on the public at large; and

6. rules discouraging waste of resources.

This is not an exhaustive list of categories, but it does include the basic areas of mining law.

Although often overlooked in discussions of mining law, it is important to remember that the mechanism for enforcing rules is an essential element of any effective legal system. The general mining laws of the United States are usually given credit for having provided an adequate legal framework for the development during the last century of a strong minerals industry in the western states. Chief among the several reasons for the success of these laws was their enforcement by the reasonably efficient and fair judicial system, territorial, state, and federal, which was extended into the western regions at an early date.

In the common law countries the source of mining law is found in statutes enacted by legislative bodies and in decisions of courts. In the United States, acts of the executive branch and administrative lawmaking are also sources of law. Courts frequently provide interpretation of statutes or common law remedies for problems never considered by legislative bodies, and consequently the courts are a prolific source of law.

Rules affecting exploration and mining, whether developed by statutory or decisional law, reflect the policy of the authority making and enforcing those rules. If the authority is a government, and if that government is responsive to the attitudes of a broad public constituency, it can be said that the rules reflect public policy. A legal framework permitting private ownership of mineral lands reflects a basic political and economic judgment that minerals should be developed by the private sector. Judgments to the contrary have been made in some countries.

A significant policy goal of governments has been to generate revenue for the support of government. This is clearly shown in the attitude of almost all governments, but it is not a uniquely contemporary goal. The interest of states in receiving a share of production can be traced to ancient times in Roman and English law. But not all policies have been so basic. The public land laws of the United States have variously served such policy goals as conservation of war materials, resettlement of war veterans, discouragement of immigration by undesirable aliens, drainage of swamps, reclamation of arid lands, prohibition of monopoly, promotion of resource exploitation, and regional development.

Mining law is dynamic, changing in response to societal needs and wants. Judicial law-making is particularly responsive to changing public attitudes. Judicial construction of mineral reservations in deeds provides an example. In several recent cases courts have adopted a narrow view of the rights conferred on the owner of a severed mineral estate. These were cases involving an attempt by the mineral owner to destroy all or part of the surface of the land in order to extract minerals by surface mining methods. Cases decided in the past have often permitted total surface destruction if the language of the grant or reservation in a deed was broad enough to clearly confer such a right on the mineral owner. In the recent cases, however, courts have prohibited such total destruction, basing their decisions on such strained theories as reconstruction of the original grantor's intentions. The real basis for the decisions is, of course, a policy judgment by the courts that strip mining should be made to pay for all of the damage it does to other resources, including surface lands which are being put to use for some other purpose by the surface owner. The mineral owner's argument that he or his predecessor in title paid for the right to destroy the surface when the mineral reservation was created is swept aside by the courts as they adjust land use conflicts in terms of today's land values. This is an oversimplified description of trends in the law of concurrent ownership of surface and mineral estates, but it does show how courts change law.

In the foregoing example the change imposed greater burdens on mining. If our national minerals policy goals were clearly stated and were supported broadly by the public, the same mechanisms for changing the law would be employed to encourage mineral-producing activities.

The dynamism of the law is counterbalanced by its adherence to precedent and its reluctance to embark upon new courses of action too quickly. This conservatism troubles social reformers, but it is soundly based in reason. The statutory and judge-made laws which comprise mineral lands use law have been developed over a period of several hundred years and can be regarded as the embodiment of society's cumulative experience with mineral land use problem-solving. This body of precedent permits us to review how problems were solved in the past and to determine whether the policy judgments adopted in solving those problems were sound. As the past slips away from us, we become presumptuous in the view that our age is the first to confront problems of great magnitude and complexity. Careful examination of legal precedents reveals that there are surprisingly few new problems in the field of mineral land use. The late nineteenth century dispute over mine tailing disposal between California placer miners and their downstream agriculturalist neighbors is a good example. The magnitude of the harm alleged in that dispute compares in size to modern disasters such as the Santa Barbara oil slick and the Torrey Canyon spill. Both the judicial and legislative branches of government, state and federal, were called upon for help in resolving the problem. The solution they adopted is profitably studied today.[1]

The Present Law Framework

A complete analysis of the present land law framework of mineral-producing activities in the United States would require a multiple-volume treatise. For the purposes of this discussion I will highlight certain of the major variables in the framework.

The development of off-shore mineral supplies by United States producers involves host-country land laws. In most cases these laws are beyond the influence

1. For a good brief account of the problem and its solution, see 3 Lindley, *Mines* § 848–853, 3d ed. (1914).

of United States policy makers. However, where land law restrictions are imposed primarily as trade barriers or in retaliation for United States restrictions, it may be possible to secure relief through bilateral agreements.

Federal land law influence on minerals production centers upon the availability of federal lands and the laws permitting disposal of such lands. The constitution places the control of public lands management and disposition in the hands of Congress, but today executive agencies such as the Departments of the Interior and Agriculture have great power over public lands.

Withdrawal of public lands from operation of mineral disposal laws has created serious problems for mineral explorers and producers. The acreage withdrawn is great, and the widespread distribution of the withdrawals has magnified their impact. The closure of public lands makes it more difficult to assure that mineral supply goals will be achieved with domestic production.

The primary federal mineral disposal acts are the general mining law which permits the location and patenting of a wide variety of minerals; the mineral leasing acts which cover oil, gas, and coal, and several nonmetallic minerals; the acquired lands leasing act; and the material sales act. Mineral producers are generally happy with the basic mining law and the mineral leasing act but feel that these laws require certain important amendments. Difficulties between mineral producers and federal agencies, exacerbated by pressures from citizen environmental groups, have made it difficult to achieve needed changes in these laws. The mining law is deficient with respect to prediscovery protection, and mine development is severely hampered by the lack of disposal laws permitting timely acquisition of lands for mining-related activities such as processing and waste disposal.

Lands containing valuable deposits of uranium and oil shale have been withdrawn from disposal, hampering private development of these resources.

The National Environmental Policy Act (NEPA) is of recent origin, but it is having an enormous impact upon public land management and disposal. This act, which is based upon sound environmental polioy goals, has become the basis for severe disruption of mineral-producing activities. It is not my purpose to argue the wisdom of NEPA or to question its review procedures, but rather to state the fact that the environmental impact statement provision of that act, and the enforcement of that provision by the courts has disrupted major mineral projects. The examples of the Alaska pipeline and the Louisiana off-shore oil and gas lease sale are well known.

It seems likely that federal land use planning laws and mined land reclamation laws will soon be added to the framework. Coastal zone management regulations are close to enactment.

State law continues to control much of the land-related activity of mining. State law controls most

relations between owners of surface and mineral estates and between owners of adjacent lands. State tax
laws, especially those which base property taxation on
the value of minable reserves or which tax the privilege
of severing minerals, have great impact upon mining,
even to the extent of discouraging development and
raising cutoff grades.

State laws also govern the disposal of state lands for
mineral development. There is a trend toward state
involvement in land use planning and zoning. Most
states now regulate mined land reclamation, a subject
of great importance to mining.

Zoning laws at the county level are beginning to
affect all mining operations. A few years ago zoning
was confined mostly to municipalities or urban coun-
ties, and the impact of zoning was usually restricted to
sand and gravel operations. Today the trend is toward
rural county and state-wide zoning. The impact on all
mining operations will soon be felt.

The systems for implementing land law disputes
have undergone considerable change in the past forty
years. The growth of the administrative law has had
a profound effect upon the land law framework. Rule-
making and adjudication functions have been turned
over to agencies charged with land-managing respon-
sibilities at both the state and federal levels. In very
recent times the administrative law system has been
under severe attack, especially from environmentalists.
There is growing disenchantment with the system, and
we may witness a return to reliance on the courts for
adjudication and enforcement functions. It would

seem that the courts are better equipped to handle many of these functions, particularly with respect to scrutiny of governmental actions, and much of the dissatisfaction with the judiciary which led to creation of administrative law has been removed in recent years by better judicial administration.

The intricate web of federal, state, and local laws and restrictions dealing with mining is further complicated by inconsistencies in the policy goals which underlie these laws, unevenness in their enforcement, and overlapping jurisdiction of the enforcement agencies. As a result, today's mineral producer encounters legal difficulty and frustration at every step from the initial exploration for a mineral deposit to ultimate development and operation. The costs of such frustration to the national economy are obvious, and it seems doubtful that the several policy goals involved are actually achieved.

Land Law and the Implementation of Mineral Policy

I have described the elements of mining law in some detail because I believe that as an institution it is well equipped to accommodate the changes which will be required if we are to meet our mineral supply goals. The present land law framework is clearly deficient, but such deficiency does not arise by reason of weaknesses in the institution of law. Rather, our present land law accurately reflects a confused public policy. It is that policy which is deficient, and it is that policy which must be changed.

Historically, mining has been regarded as highly beneficial to society. The products of mining have been considered essential to our well-being. Miners have often enjoyed a favored legal status because of their contributions to the common good.

The other authors in this volume have adequately described the mineral requirements of the United States during the period 1975–2000. If we are to meet these requirements a systematic approach to policy formulation and execution must be adopted at once. New mineral policies must be broadly communicated to all levels of government and the public, and both public and private sector resources must be employed in carrying them out. I believe that it will be necessary to re-establish the attitude that mineral production is of special importance to the nation and to once again give certain preferences to mineral-producing activities. In particular, the land law framework must be changed to eliminate unnecessary constraints on exploration and mining.

In this regard it is disappointing that the Department of the Interior's first annual report on a national minerals policy, required of it by the National Minerals Policy Act, devoted so little attention to the land law framework as an element of minerals policy. That report included just one small section dealing with problems of withdrawal of lands from operation of mineral disposal laws. It is to be hoped that, as the department gains sophistication in its analysis of minerals policy problems, it will devote attention to public lands problems.

Formulation of minerals policy and land law changes will require that some tough choices be made. It will be particularly difficult for many people to accept the concept of dominant use, but the adoption of some type of dominant use theory is inevitable if minerals are to be produced in the required quantities. The Public Land Law Review Commission came to this conclusion, recommending that "mineral exploration and development should have a preference over some or all other uses on much of our public lands."

In urban areas where sand and gravel supplies are becoming scarce, there is an emerging comprehension of the need for protection of sand and gravel deposits from urban development. In these cases, however, the adjustment of the land use conflict is time related, and the time is short. Developers must wait only a short time before the deposit is harvested and the land is available for other use. The more difficult question will arise soon, as land use planners become aware of the mineral provinces that contain the lands most likely to be underlain by undiscovered deposits of metallic ores. Will the planners recognize that mining should be a dominant use, and protect these areas from uses such as recreational housing? I submit that as a nation we will make these difficult choices in favor of minerals production, but not without an agonizing adjustment of attitudes.

It also seems likely that the United States will soon be faced with difficult decisions in the environmental field. It may seem early to predict a retreat from environmentally prompted extremes, but I believe that this nation will soon be forced to adopt environmental standards which are "economically tolerable," and to limit the National Environmental Policy Act review of federal actions to those actions which are truly *"major federal actions."* The United States will, if it is to support its projected minerals-based economy, be forced in many cases to choose between preservation and conservation of environmental resources.

If a national minerals policy is clearly announced and if that policy is supported by the public, much of the change in law required will be accomplished without sweeping legislative change. The American system is flexible enough to accommodate such change without radical law "reform." However, I would like to point out two problem areas that could interfere with orderly change in the land law framework of increasing mineral production.

In order to free mining from unnecessary restrictions, legislators must avoid the temptation to use land tenure laws to accomplish "other" policy goals. The prime example today is the attempt to make land tenure under the mining and mineral leasing laws conditional upon compliance with environmental regulations. The most basic requirement of mineral ownership is security of tenure. Conditions threatening such security diminish the value of a mineral property and make it difficult to secure financing for development. There are certainly other ways to achieve environmental goals without threatening tenure. The mining industry has consistently expressed support for reasonable environmental laws of general application, but it has opposed attempts to use interference with tenure as a sanction to enforce environmental regulations. The industry also points out that land tenure laws, which are the basis for huge investments, tend to be resistant to amendment and are therefore poor vehicles for social legislation, which must be flexibly drawn and repeatedly amended. Fossilizing environmental control requirements in inflexible property laws is not good environmental policy.

Congress has tried in the past to achieve many different policy goals by engrafting onto public land laws various limitations. Most of these attempts have failed to secure the ancillary goals, and some may have been inimical to the underlying economic goals of the laws. For example, it is interesting to note that this nation's public land laws still contain restrictive provisions relating to direct and indirect ownership by aliens of interests in public lands minerals, and that the Public Land Law Review Commission was compelled to state that "in view of the substantial overseas commercial and investment interests of United States corporations and individuals, we believe existing restrictions on aliens should be removed except when required by explicit foreign policy considerations of general applicability to transactions of aliens."

Finally, and most important, we must find a way to eliminate the confusion and conflict that now pervades the land law framework. It will not be possible for this nation to secure its mineral supply goals if the various agencies of government continue to follow independent and disjointed policies and compete with each other for jurisdiction, funds, and public attention.

Ian MacGregor, chairman of American Metal Climax, Inc., in a recent paper on trade policy and the role of United States mineral producers in international mineral development, made these comments on the relationship of the United States government and American industry:

Traditionally contacts between business and government have been stiff and formal. It has been a dialogue between the regulator and the regulated. Government has been fragmented. Agencies created at different times in the nation's history, in response to different public demands, follow their own mandate, often without much communication with each other.

In some ways this fragmentation may be useful. It is certainly compatible with our notions of independent business and a system of checks and balances in government. But a government made up of separate legislative and administrative branches, the latter being further subdivided into a series of semi-autonomous units, each pursuing its own policy goals, often heedless of other national needs, is not well equipped to identify our nation's interests in today's complicated and highly competitive world marketplace.

The fragmentation of which I speak is not the result of conscious design, and the lack of communication between business and government does not reflect a planned estrangement of these important institutions. Rather, it is a product of our nation's history and its political and economic traditions. In many ways our uninterrupted growth and development as a nation has permitted us to refine our institutions and to capitalize on past experience. But the past also encumbers us with some excess baggage which we sometimes find politically difficult to unload.

The fragmentation of governmental policy described by Mr. MacGregor is a prominent feature of current American land law framework.

In summary I suggest that the legal institutions of the United States, and particularly mineral lands use law, are flexible enough to accommodate a shift toward more favorable treatment of mineral production. It is our political institutions that require change in order to secure adoption and execution of mineral policy consistent with our projected national needs. If policymakers draw upon the collective wisdom contained in that body of precedent we refer to as "mining law," retaining the good and rejecting the bad, the nation should be able to achieve its mineral supply goals during the period 1975–2000.

8 *John Drew Ridge*

Minerals
from Abroad:
The Changing Scene

United States Dependence on Mineral Imports

If the United States is to obtain the mineral
raw materials needed to sustain a continually increasing standard of living, it must import more and more
of such things from abroad. Every year United States
consumption of mineral raw materials per capita is
larger, and each year less of the increase in nearly
every mineral material we use is provided by our own
mines. Despite the huge tonnages of mineral raw materials obtained from mines within our own borders,
the gap between what we produce and what we need
is getting steadily greater for almost all categories of
mineral products. This means that we are increasingly
dependent on supplies from abroad and will have to
import a larger share of our mineral requirements each
year even if our standard of living does not go up. A
standard of living in any modern industrial nation,
however, cannot stand still. It must go up or down, and
we continue to assume that ours will go up. Despite

wishful thinking abour zero population growth, it is
unlikely that this goal will be reached in this country
in this century. Demand will, therefore, increase not
only due to the existence of more people but also due
to their demands for more goods than each of us uses
today. The problem, then, is not whether demand will
grow but whether the United States can satisfy the
greater demand engendered by more people who want
more things or whether we will have to reduce our standard of living because we cannot produce enough to
supply each member of our expanding population at
an increasing rate. It is possible that America's continually improving technology will enable us to meet an
increased demand for goods with a smaller amount of
metal raw materials per person per year, but this
improvement is unlikely to offset completely the
increase in demand caused by the nation's growth in
population and its citizens' desire for more and more
complex goods. Further, as more mineral raw materials have to be imported from nations that will
steadily demand a higher net return for their products,
the costs of mineral materials will rise more rapidly
than will the ability of America and Americans to pay
for them, thus putting an even greater burden on
technology.

It is obvious to anyone who has considered the
matter that it is impossible to furnish the entire population of the world with goods and services at the rates
they are consumed in this country. To supply all people
with what the United States has would require more

mines, mills, factories, and distribution systems than could possibly be developed with the resources—natural, manufactured, and trained human—that the world now contains. Thus, if we are to maintain our position of mineral use, relative to the rest of the world, we will have to remain as the greatest per capita users of mineral raw materials in the world. As world population probably will grow more rapidly than America's, the disparity between this country's way of life and that of many of the nations of the world will grow rather than diminish. This seems to follow even though some evidence has been presented in chapter 2 of this book that world mineral consumption is increasing more rapidly than that of the United States. What we do not know is this: will the people of the developing nations, on whom we will depend more and more for the minerals we need and from whom we must buy a greater and greater percentage of our mineral needs, continue to provide us with them in increasing quantities?

Although the United States makes up only about 6 percent of the people of the world, it consumes a much higher percentage of the world's mineral production than does any other nation. How much higher is a matter of some uncertainty, but, if the other nations of the world wish to consume more per capita than they do now, their additional needs must be met mainly at our expense. Even though we still produce the major portion of our mineral requirements within our own borders, our dependence on foreign sources will continue to grow. But this does not mean that we will be able to get what we want unless we can afford to outbid other potential consumers of the same mineral goods. If we outbid our competitors, we will send more of our available funds for minerals than we do now. This will mean, essentially, that we will have to allow our standard of living to lower so that the rest of the world can improve theirs. To the extent that the United States can make up for these increased charges for minerals by becoming more efficient in its use of them, we can reduce this drop in our living standard, but some decrease there almost certainly will be.

The major factor in the country's favor for the next twenty-five years, the period we are considering here, is that nations that produce large quantities of mineral raw materials, far above their own requirements or abilities to consume them, must sell their products abroad or face financial disaster, with all the political instability and human misery that that entails. Chile, for example, must sell copper to the United States no matter what the opinion of its ruling junta may be of the government of the United States. Cuba must sell mineral raw materials to the Soviet Union and the Mainland Chinese, since the United States will not permit its citizens to buy from Cuba. If political ties with the major communist nations were lacking, Cuba would have to develop them to prevent financial chaos. The same reasoning, of course, applies equally to Cuba's agricultural products, principally sugar. Thus, nations with large surpluses of agricultural or mineral products must sell them abroad (witness our recent contract to supply wheat to the Russians). If Europe needs oil from the Middle East, that area also needs the income from the oil to maintain even its present low standard of living. The problem for the European nations, of course, is that the nations of the Middle East can get along for a while without oil income because their citizens are used to hardship, but the industrial nations of Europe cannot get along without oil for a single day.

Sources of Mineral Imports for the United States

The sample situations just outlined demonstrate that America's best hedge against the cutoff of its raw materials supplies is to have alternate sources to which it can turn if one area or nation decides to withhold its mineral products from the country or if the United States refuses to take them. In some ways this appears difficult to do. The number of areas in which we can obtain concessions to prospect for mineral raw materials and later mine any discoveries we may make grows smaller with the passing of every year. In one way or another, through overt expropriation (or, more euphemistically, nationalization), tightened requirements for profit-sharing, control of mining companies by host-country nationals, greater taxation, increased wages, and higher royalties, it is becoming increasingly difficult and expensive to find and mine the mineral raw materials the United States needs from other lands. In one year a company may drop from owning 100 percent of a highly profitable investment to owning 49 percent in one that, because it is now controlled by a government lacking in training and experience in exploration and mining ventures, is less profitable. Add to this that the purchase price often is paid in bonds issued by the host country, payable over a long period in the steadily inflating currency of the host country and subject to cancellation if an even more extreme regime should achieve power, and the prospect for investment in mining abroad seems less than attractive. Furthermore, next year, or five years hence, or within whatever period of time the internal politics of the host countries mandate, the foreign company that owned 49 percent of this enterprise may find itself owning nothing at all and its treasury holding an even larger number, I would hardly say value, of host-country bonds. The operation being now 100 percent under the control of the host country, and the supervision that cannot be supplied by host-country nationals being now carried out by foreign nationals on short-term contracts, it can be no surprise if production drops and inflation in the host country is spurred. This is particularly likely if the host-country economy is not viable without the maximum possible infusions of foreign exchange provided by the sale of the mineral raw materials. Suppose that the tonnage of the mineral material produced in a developing country, copper for example, drops by only 15 percent. This 15 percent may well make the difference between the host country's being able to meet its bills from abroad and being unable to do so. Such a situation means that the host country will have, at best, less consumer goods to sell to its citizens, that prices will be forced up, that wages will have to follow, and that inflation will be generated at an even faster rate. Under these conditions, the bonds issued to the foreign mining company for 100 percent of its holdings lose value even more rapidly than those issued a few years previously for 49 percent of its property.

For a mining company of an industrialized nation in western Europe or the western hemisphere, one solution to the problem of nationalization that seems quite obvious, at least at first glance, is to concentrate its activities in nations having governments that welcome foreign investment in, and ownership of, mineral properties and have records of political stability that would seem to guarantee the safety of investment for years to come. Politically stable nations in which such companies can expect to obtain large-scale concessions for both exploration and mining are limited. Basically, these are Canada, Australia, and South Africa, but Rhodesia is geologically attractive and politically stable enough, whether or not the problems created by that country's unilateral declaration of independence from Great Britain are solved in the near future.

The countries next most attractive to foreign mining companies are nations such as Mexico and Brazil that are politically stable and geologically favorable but require that 51 percent of any exploiting company be owned by their citizens. In addition, exploration and mining concessions can be obtained in such countries as Ireland, Iran, and Spain. In these countries the governments seem quite firmly fixed in power, with the government of Iran being the least firm and that of Ireland the most firm, so long as Northern Ireland is not included with the Irish Republic.

Turkey is in a position all its own. Concessions can be obtained there and the mineral potential is high, but the difficulties of dealing with Turkish officials and the problems engendered by a rather turbulent minority that opposes the established regime make Turkey less favorable than the three nations just mentioned.

The considerable expanse of Indonesia (its total land area is nearly as large as that of Iran) and the apparent stability of the Suharto regime, plus the definite interest of the present government in encouraging the investment of foreign capital in mining operations, make it a nation to be given serious consideration by American mining companies.

The major mining countries on the west coast of South America no longer offer areas in which American companies can expect to obtain concessions of the type that once permitted Kennecott and Anaconda to operate so profitably in Chile and that still allow Cerro to work its mines in Peru. Both these countries, however, have been able to borrow money from banks in other nations, and only time will show whether these investments were worthwhile or not. As has been pointed out for Chile, neither Chile nor Peru can operate its mines profitably unless it sells the products, at least in major part, to the United States, but such sales normally will be as small as the governments in question can make them.

The present government of Argentina is attempting to attract foreign capital and experience for exploration in the Argentinian Andes, but some question also exists about the stability of the Argentina regime. Granting stability, nevertheless, considerable numbers of strong nationalists in the country are arguing that the mineral resources of the nation should not pass out of its control and that Argentinian capital should control, and state enterprise should operate, any mining venture.

Venezuela has already indicated what it intends to do with the concessions granted to American oil and iron mining companies when they expire; it will take them over lock, stock, and barrel and will pay compensation based on a value determined by Venezuelan officials. The chances that new concessions will be granted are negligible.

Appreciable opportunities for mineral exploration and mining are available in the West Indian Islands and Central America. Certainly the relationships between American mining firms and the government in Jamaica show how successful operations can be in this region, although the slow progress toward the opening of the porphyry-type copper deposits found in Puerto Rico is not encouraging.

Southeast Asia is hardly attractive to any American company desirous of investing in exploration and mining abroad. The region is highly favorable geologically, but the probabilities are overwhelming that the entire area will pass under communist domination within a few months after the United States has withdrawn all support from the present governments of South Vietnam, Laos, and Cambodia. Thailand probably will not last long after the other countries have come under communist control.

The Philippine Republic is another country in which geology is far more favorable than political conditions. Although several American firms are operating there, through Philippine-controlled subsidiaries that must be owned at least 60 percent by Philippine nationals, the presence of a native guerrilla movement descended from the HUKS of World War II makes the area appreciably less attractive than its geology would suggest.

In India the government is willing to accept American assistance, such as that provided on so elaborate a scale by the Ford Foundation, but a regime with such strong socialistic convictions is not going to allow American geologists to prospect on its territory as agents of foreign companies. China, of course, is closed to such activities and will continue to be so for a long time in the future, despite President Nixon's visit to that country in 1972. With the exception of southern Africa and Liberia and possibly Nigeria and Ethiopia (in which no large-scale mineral deposits have been found despite efforts extending back into the nineteenth century), the major countries of the African continent have no intention of granting total-ownership concessions to American firms for exploration and exploitation. Similarly, the Moslem countries, with limited exceptions in Iran and Turkey, will not allow, much less encourage, American participation in mineral exploration and production on a concession basis. Most of the European countries are perfectly capable of doing their own exploration and of raising the funds necessary to put into operation any discoveries made within their borders. Thus, the greater portion of the earth's surface appears to be closed to American mineral operations in the sense of completely controlled concessions, and it will continue to be so until the end of this century at the very least.

From this discussion it appears obvious that one way in which the United States might insure that its mining operations abroad could be continued in the manner of the past seventy-five years is to concentrate its exploration and extraction operations in those countries in which concessions still can be obtained and in which the laws are not likely to be changed drastically for the remainder of this century at least. The list of major countries in this category is not long—Canada, Australia, South Africa, and Rhodesia.

In Canada the recent reforms in the laws governing personal and corporate taxes suggest that the profitability of the mining industry in Canada will drop.

The Mining Association of Canada claims that the new act already has reduced exploration and development activities in that country and ultimately will seriously damage both the mining industry and the entire national industrial complex. Certainly exploration and development dropped off in the last few months of 1971, but how much of this decline was due to the new laws and how much to the recession in the world demand for metals is not clear. The full impact of these new regulations is difficult to assess at this time, but they certainly add to the complexities of doing business in Canada and force American firms to add them to the variables that they must consider in choosing between Canada and other countries as sites for the expenditure of funds.

As things now stand, Australia is probably the most satisfactory country, from a political and an economic standpoint, for mineral exploration and exploitation by an American mining firm. It is only in the last year that the profitability of exploitative operations on that continent has been called into question. The most serious problem, and it is one that is particularly so for firms that have deposits that are just entering or about to enter into the production stage, is that world demand for mineral materials of many kinds has dropped drastically since the latter part of 1970. Since the Japanese are Australia's best customers for minerals, the recession that has hit that island has been felt directly in Australia. The attitude of Australia toward foreign investment in its mining sector may well become less favorable as a result of the triumph of the Labor party in the December 1972 election. Nevertheless, despite a variety of economic and financial problems, Australia today seems the most favorable area in the world for the investment of the funds and skills of American mining companies.

What is the situation in South Africa? To anyone not particularly sensitive to the overtones of race, both black versus white and English-speaking white versus Afrikaans-speaking white, the situation in that country may seem almost ideal for exploration and exploitation. Some recent events (the decisive defeat of the extreme Nationalists in the April 1970 elections, for example,) suggest that the Afrikaans-oriented members of the population have decided to live with the world as they find it, accepting industrialization as the price that must be paid for an expanding economy. Beyond the problems of race, however, South Africa must face one that may be even more serious in scope and more current in time; this is the shortage of water in much of the country, a shortage that is growing worse as one low-rainfall year succeeds another (although 1972 may be an exception to this pattern). A reasonable solution to all the political problems of South Africa would be meaningless if the needs for water are not met. For an industry that uses as much water as does mining, this problem could cause serious difficulties.

What do all these problems mean to the firm that is considering major expenditures in South Africa for

exploration for, or production from, mineral properties? For any operation that can be completed, from
the first preliminary survey to the mining of the last
pound of ore from the deposit, within twenty-five
years, the risks are probably not too great, although
a continuation of the drought may reduce this length
of safe time by ten years. But beyond the twenty-five-
year period, the South Africans may well have to face
armed conflict with the allied nations of central Africa.

Rhodesia: A Special Case

The political problems to be weighed by any firm
intending, as soon as United States law permits, to
invest time and money in Rhodesia are at once both
greater and less than those in South Africa. On the
one hand, the white community is far more homogeneous than that in South Africa; for all practical
purposes, no division exists between English- and
Afrikaans-speaking groups. On the other hand, the
whites are outnumbered to an even greater degree
(as much as twenty to one) by the blacks than are the
whites in South Africa. What does concern us, however, moral considerations aside, is the effect of the
Rhodesian declaration of independence and the
United Nations sanctions that followed on American
access to mineral raw materials (now mainly chromite)
from that country.

During the period in which the United Nations
sanctions have been in effect, a very real amount of
mineral exploration has been done in Rhodesia.

Apparently there has been considerable success in finding materials other than chromite. But the United States has had little to do with this exploration, and we will have little claim on the materials produced in the future from the mines that almost certainly will be developed from these discoveries.

Various Alternative Solutions

As an alternative to obtaining the mineral raw materials the United States needs only from the four countries just discussed, we might meet our mineral needs somewhat in the manner of the Japanese, that is, by carrying out essentially no exploration or mining ourselves but buying what we need from countries that do. If, however, the United States were to follow the example of the Japanese and do little active exploration outside its own borders, the number of new mineral deposits found outside Russia and its satellites would be far smaller than it is now. To some extent, the discovery gap that would be left if the United States removed itself from mineral exploration might be taken up by other western countries. They, however, face the same problems in obtaining concessions that we do, and, if we adopted the Japanese policy, our western peers might do the same. This approach would leave the developing nations with the burden of accomplishing their own exploration if they wanted to continue in, or enter, the mineral field. But they normally lack the necessary personnel and the funds to embark on exploration programs themselves,

so they must seek elsewhere than in western Europe or
North America for the trained men and money (or
credit) to get the job done. Any developing nation
might hire countries in eastern Europe or eastern
Asia to do its exploration. Several emerging nations,
Algeria, Iran, and Nigeria, for example, have gone
behind the Iron Curtain to such organizations as
Polservice (Poland) for the exploration skills they
need. Not only Russia but also most of the nations
satellite to it will provide such teams for a price; this
price may not be in cash (often not available anyway)
but in political and economic concessions. So will the
Japanese and the Chinese. Granted a minable deposit
is found by such a visiting organization, it will be in a
good position, set out in the exploration contract, to
sell equipment and technical skills to the country in
question for the exploitation phase and perhaps to
lend money or, more likely, provide credit for that
phase. The nations providing the exploratory skills,
therefore, are in a position to make a profit on exploration, production, and perhaps the marketing phases
of the operation. Although the emerging nation may
set up its own marketing firm or join with other similar
nations in creating a market cooperative, it may lack
the experience and financial skill to do so and may
wish to delegate this function to some other nation or
organization it can trust. Before, however, the marketing phase can be reached, the funds necessary for the
production phase must be acquired. If, for example,
huge metallic mineral deposits were to be found in the

Precambrian shield areas of Saudi Arabia, the oil-revenue-rich government of that country easily could provide the money needed to put the deposit into production. On the other hand, if a similar deposit were to be found in like terrain in Ethiopia, the funds required hardly could be provided in that country. The money, therefore, must be borrowed. For this service, the World Bank (International Bank for Reconstruction and Development) might be approached. This application would be successful only if the bank were convinced that the deposit was worth working and that markets existed for the product at prices that would guarantee a profit both to the lending agency and to the nation concerned. If either the Soviets or the Chinese wished to gain political credit in the country in question or needed the raw material being produced, one of them, perhaps both, might offer to provide, singly but not together, the necessary funds or credit. Since no truth-in-lending act has been passed that is applicable at the international level, the cost of the loan might be appreciably greater than one obtained by bargaining in the open market, but the emerging nations, in return for previous favors, may have drastically limited their options.

So, both the United States and the developing nations would suffer if we were to adopt the Japanese system of seldom exploring (except under contract) and of investing in foreign mining operations only occasionally. Instead, another alternative exists that I think the United States would be well advised to adopt. All American companies (or combinations of them) that wish to be involved in foreign exploration and production should be willing to sell their exploration and production services to the nations wanting and needing them. This approach is now largely preferable to attempting to obtain concessions for such operations. The first problem to be met would be that of obtaining funds for exploration. The country in which the exploration is to be done might be able to provide the funds from its own resources, or it might obtain loans from the World Bank or large commercial banks (or groups of them). As another alternative, the exploration consortium might be willing to lend the needed money itself if the probability that the investment would be repaid is high enough. As a further charge for carrying out the exploration phase, the firm or firms involved could insist on being granted a service contract to mine and process any workable ore bodies found in the exploration phase. They could also insist on an arrangement for marketing the product, or, at least, on the right to buy all or an appreciable fraction of it at reasonable prices. Under such a system the American firm or firms involved would be in a position to make a profit on the first three phases of the operation (exploration, production, and marketing) and, even more important, on the later fabrication of the metal into goods to be sold at home and abroad. By concentrating most of its available funds into this fourth phase it could invest its funds to greatest economic advantage.

Obviously such an approach would require a complete reorientation of effort by American firms, who are quite reasonably biased in favor of a system that has worked so well for so long. No company can be expected to decide that it will give up concession exploration overnight and engage in nothing other than selling its services to foreign governments. If nothing else, many foreign governments might look on such a change of heart as a subtle plan to win control over their mineral resources under the guise of a beneficent lack of self interest. Rather, any company deciding to undertake prospecting under contract should do so gradually. Certainly, the experience of numerous consulting firms, of the United Nations in Sonora, Argentina, and Panama, and of several East European nations in countries from Mongolia to Nigeria shows what could be done if larger resources were made available and a greater variety and depth of experience applied to the problem of ore-finding. Such work by American firms would have to be done in direct competition with socialistic countries already well established in the business and in indirect competition with the United Nations. Neither these countries nor the United Nations, however, has the personnel and experience available to American organizations.

Recently, many American mining companies have drastically reduced or even completely eliminated their exploration departments. The rationale behind such acts is to lower costs of activities on which current operations least depend at a time when demand and prices are down and unit costs are up. This is not a new phenomenon in the American mining industry, but it is one that has had and will have serious deleterious effects on the efficiency and effectiveness of its exploration functions. When demand again picks up, and reserves must be increased, the companies will be (and have been) surprised that long lines of experienced exploration geologists do not come trooping to their doors to engage again in the highly insecure profession of looking for ore deposits.

Some companies, while cutting down on United States exploration, have increased their activities in that field abroad; this makes possible the transfer of personnel not needed in this country to foreign work. This does not, however, by any means constitute a complete answer to the problem of maintaining a good exploration department through good economic times and bad. If, however, these firms were well entrenched in the business of exploration, under contract, of developing nations, they would have another way in which they could use their exploration geologists and help keep their exploration groups intact. At such times, it might well be worthwhile to sell the services of the companies' exploration teams at rates that would do little more than pay actual operating costs but would not damage the cash-flow positions of the companies in question. Further, such bargain rates would help gain an entrée into countries where later appreciably more lucrative agreements to mine the ore discovered by low-rate exploration might well be obtained. This is particularly true because the cyclical nature of the mining business suggests that economic conditions, whatever they may be at the moment an exploration is undertaken, will be different when the work is completed. Since we are now near an economic low-point, conditions by the time an exploration project begun today is finished almost certainly will be better.

Other arguments for retaining exploration teams of proved efficiency can be put forward, such as the better terms that can be obtained for concession exploration in times of poor economic conditions, but they are not pertinent to the argument being developed here and will not be considered further.

Production under Contract

Beyond the prospecting phase, granted that discoveries have resulted from it, the production phase offers broad opportunities for profit with far less risk of loss than under the concession system. The hundreds of millions of dollars required for the mining of a huge porphyry copper deposit would no longer be the responsibility of the company or consortium bidding on the conduct of the mining phase of the enterprise. Instead it would be up to the host government to find the money. Backed by favorable feasibility studies prepared by the consortium, the host government should be able to enter the world's financial markets with reasonable chances of obtaining the funds required. Under certain conditions, the consortium might wish to invest some of its own money in the enterprise. This would, of course, be in the form of a loan to the host government; on such a loan the rate of interest might be less than achieved on equal sums invested in a wholly owned mining venture, but also the risks of loss would be far less. No consortium in its right mind (if consortia can be thought to have minds) would lend its money to a country of dubious financial stability. Using Saudi Arabia again as an example, little risk would attend the lending of money for a mining venture to a country with such an enviable financial position, a position resulting from its oil revenues. On the other hand, the present condition of the copper market might well give an American organization second thoughts about lending money for the development of a new, large copper deposit in Zambia even if the organization were to have direction of actual mining. Thus, the approach suggested here is not a complete answer to the problems of using the prospecting and mining skills of an American company or consortium in foreign ventures. Profits would not be guaranteed by this method, nor would they be potentially as large as under the system now favored by such organizations. The risks involved, however, would be appreciably reduced from what they currently are, particularly if an organization such as the Overseas Private Investment Corporation (OPIC) were to provide insurance against loss at reasonable rates. The present rules under which OPIC operates would have to be revised if a sufficiently firm guarantee were to be provided against loss, but the principles of such insurance have already been established.

Transition from Concession to Contract

The transition phase from concession to service-contract operations would have to be accomplished in steps. Most American firms are engaged in foreign operations in what they consider to be politically reliable nations and have no desire whatsoever to cancel these in midstream. In fact, it is to be expected that work on the present basis will continue in these countries as long as the laws of the country in question offer a fair chance of a profit. But it would seem reasonable for most consortia to give serious consideration to offering their services to nations that wish to control what is done with their national resources. Once a few successful operations of this sort have been carried out, the resulting word-of-mouth advertising will be sufficient to bring in additional clients.

Consortia would have to face requirements of the host countries that native personnel be trained in the various techniques of the exploratory and exploitative phases. Such requirements are only natural. However, so many people would have to be trained in so many parts of the world that it would be a long time before the consortia worked themselves out of opportunities for further work abroad.

Advantages of the Contract System for the United States

Thus far, I have said little about the advantages that would accrue to this country in general if such a system of service-contract operations were adopted by American firms, but such advantages would be quite real. Any examination of the statistics of the mineral raw material trade of the United States will show that we are not self-sufficient in most of the metal materials we need in our industrial activity and must get them from outside our own boundaries. To get these materials for ourselves in the quantities we require and at reasonable prices, we must be able to buy wherever we like at whatever times we wish. This means that we cannot diminish our options by saying that we will have nothing more to do with a nation because that nation has nationalized its own resources. When the Bolivians nationalized their tin mines in the early 1950s, the same firms that had owned the mines continued to smelt the tin concentrates, on the theory that profits derived from the part of the processes needed to produce tin metal were better than no profits at all. In short, we must be able to channel our resources and knowledge of how to do things into those areas of industrial activity that are open to us, and we must not refuse to do things in new ways merely because we have not done them in those ways before. If we must concentrate to a greater extent on final processing and fabrication and reduce the amount of resources we put into the earlier stages of the recovery of metal materials from their ores, we must adapt to this. Most American firms have adapted to the world as it is. If the United States mining industry does not, it will have lost its position as an industrial leader and must accept the reduced influence in world mining affairs that such a loss would entail. But American miners do not have to sit back and watch the world change without them; they can change with it and do so at a faster pace than any other mining nation.

Conclusion: Competition a Necessity

From this it follows that the best interests of the United States will be served by encouraging and promoting vigorously competitive markets for minerals. This is best done by doing those things that will make minerals available in abundant supply from a variety of producers and will insure that minerals are sold to the highest bidders, without favor, from mines that have been found and developed with all due speed.

We must remember that ownership abroad in these times is an insecure basis on which to build a metal mining business. What guarantees a sound supply of mineral raw materials is the competition between source nations that develops when several nations have the same material to sell and must do so in commercial rivalry with each other. This situation will obtain most frequently and have the longest duration if we aid in finding and exploiting mineral resources without committing large shares of our own financial resources to do so. So, instead of seeking ownership, we should seek to sell our services, which are more valuable and effective than those of any other country, and then buy the products of the operations we have made available through our initiative and ability in exploration and exploitation. Once we have secured our supplies of raw materials, our survival as an industrial nation depends on what we do with them. If we fashion them into products that can be sold at home and abroad in competition with those of any other nation, we will continue to prosper. The alternative we must be certain we never have to face.

Index

153

DATE DUE

e 15 '76			